ROUTLEDGE LIBRARY EDITIONS: ENERGY RESOURCES

Volume 5

T0203930

COAL IN BRITAIN

COAL IN BRITAIN

GERALD MANNERS

Routledge
Taylor & Francis Group

LONDON AND NEW YORK

First published in 1981 by Allen & Unwin

This edition first published in 2019
by Routledge
2 Park Square, Milton Park, Abingdon, Oxon OX14 4RN

and by Routledge
52 Vanderbilt Avenue, New York, NY 10017

Routledge is an imprint of the Taylor & Francis Group, an informa business

© 1981 G. Manners

British Library Cataloguing in Publication Data
A catalogue record for this book is available from the British Library

ISBN: 978-0-367-23168-2 (Set)
ISBN: 978-0-429-27857-0 (Set) (ebk)
ISBN: 978-0-367-23115-6 (Volume 5) (hbk)
ISBN: 978-0-367-23122-4 (Volume 5) (pbk)
ISBN: 978-0-429-27835-8 (Volume 5) (ebk)

Publisher's Note
The publisher has gone to great lengths to ensure the quality of this reprint but points out that some imperfections in the original copies may be apparent.

Disclaimer
The publisher has made every effort to trace copyright holders and would welcome correspondence from those they have been unable to trace.

Coal in Britain

Gerald Manners

London
GEORGE ALLEN & UNWIN
Boston   Sydney

First published in 1981

GEORGE ALLEN & UNWIN LTD
40 Museum Street, London WC1A 1LU

© G. Manners, 1981

British Library Cataloguing in Publication Data

Manners, Gerald
  Coal in Britain.
  1. Coal — Great Britain
  I. Title
  333.8'22   HD 9551.9

ISBN 0-04-333018-5
ISBN 0-04-333019-3 Pbk

Typeset in 10 on 13 point Times by Red Lion Setters, London
and printed in Great Britain
by Richard Clay (The Chaucer Press) Ltd,
Bungay, Suffolk.

For Anne

Foreword

The Resource Management Series reflects the view that academic research should play a crucial role in developing informed policies on resource use. In particular, research is required to establish the objective need for resource developments and to assess their physical and socio-economic implications. Without it, society may continue to choose those resource-use options that have easily definable, short-term political or economic advantages but which fail to take account of their longer term environmental and economic consequences. By the same token, academic investigation must be based on a sound understanding of the legal, administrative, economic and political contexts within which resource management decisions are made. Otherwise, research results will remain unread by policy makers and will not be incorporated into their decision-making processes.

The *Series* has been planned as an interdisciplinary vehicle for major contributions from scholars and practitioners with a wide variety of academic backgrounds. Each book reflects the individual views and experiences of the authors: no attempts has been made to impose a standard series style or length, such matters being more properly determined by the subject matter and the authors. All the books are, however, bound together by a common concern to improve our understanding of resource management policies, and all are based on substantial research or practical management experience.

For many years, Professor Manners has been a leading analyst of and commentator on British energy policy. In previously published work, he has analysed the competitive relationships between alternative energy sources and has highlighted the political and economic problems involved in the development of a consistent fuel policy. In his book for the *Series*, *Coal in Britain*, Professor Manners focusses his attention on the capital investment programme of the National Coal Board, appraising the extent to which it is compatible with projected total energy demands and with the market share that coal can realistically hope to capture.

Since the so-called oil crisis of the early 1970s, the National Coal Board has embarked upon a massive investment programme. This is to enable increased production from existing mines and to develop new mining centres such as those in the Vale of York at Selby and, more recently, in the controversial Vale of Belvoir field in Leicestershire. The investment programme presumes that coal's competitive position within the national fuel economy will improve markedly and so reverse the decline in demand for coal that has been so evident during the last two decades. Professor Manners argues that this presumption is more a statement of the National Coal Board's aspirations than a view developed from an objective assessment of the variables determining demand. Only by exaggerating the size of the total energy market and by making over-optimistic estimates of coal's likely market share has the NCB been able to justify the investment programme. The dangers involved in the premature investment of scarce capital resources, and the problems that would result from locking the economy into a highly inflexible energy development path are clearly brought out in the book.

Professor Manners's scholarly and detailed treatment of one of the crucial questions facing the country's energy planners should command the attention of all those interested in energy policy. It should become an important source of reference for students and, hopefully, will be read by all engineers and administrators responsible for the future development of our energy resources.

RICHARD MUNTON AND JUDITH REES
January 1981

Preface

In many respects this study is a child of the 1979–80 Vale of Belvoir Inquiry, which was established by the Secretary of State for the Environment to examine the proposal of the National Coal Board to open up three new coal mines in North East Leicestershire. The Board's plans met with considerable opposition, not least from the Alliance, a broad based coalition of interests which included the Vale of Belvoir Parish Councils' Committee, the Vale of Belvoir Protection Group and the National Farmers' Union. They retained counsel and a number of expert witnesses to appear on their behalf. Amongst them was the author of this study who gave evidence to the Inquiry on the issue of need – the question of whether the future demand for coal and alternative mining opportunities justified the environmental changes that coal mining would inevitably bring to a relatively unspoilt part of England.

Evidence given to planning inquiries is usually afforded only brief public exposure through reports in the press. The proofs of evidence and the transcripts of the proceedings very quickly become inaccessible to most students and the wider public. It was a desire to give wider circulation to one of the debates that was exposed at the Vale of Belvoir Inquiry, therefore, that prompted the preparation of this study which draws attention to one of the unresolved questions underlying national energy policy.

All the figures were drawn in the Cartographic Unit of the Department of Geography at University College London by Christine Hill and Sarah Skinner.

GERALD MANNERS
November 1980

Contents

List of tables

1 The plans of the National Coal Board

In response to the transformed energy situation of the early 1970s, the National Coal Board (NCB) for some years has been committed to an outstandingly large capital investment programme designed first to stabilise, and subsequently to increase substantially, production in both existing pits and new mines. The NCB's initial *Plan for coal* (1974) called for production to be raised from the 130 million tonnes that had been won in 1973, to between 135 and 150 million tonnes in 1985. Then in 1977 the Tripartite group, representing the government, the Coal Board and the National Union of Mineworkers (NUM), proposed in *Coal for the future* that the production target for the year 2000 should be 170 million tonnes, of which at least 150 million tonnes would be deep mined. Both of these objectives were firmly embraced in the Labour Government's Green Paper on *Energy policy* (Secretary of State for Energy 1978), although the precise route that the NCB has sought to follow in order to reach these objectives has changed over the years.

Plan for coal sought to reach the 1985 target of 120 million tonnes of deep mined coal (plus 15 million tonnes opencast) by means of 20 million tonnes from new mines, 63 million tonnes from existing pits with major modernisation investments, and 37 million tonnes from older and smaller pits. By 1977, however, it was clear that the process of planning for and sinking new mines would take much longer than had been estimated a few years earlier, and that the most such facilities could be expected to yield by the middle 1980s would be 10 million tonnes. Fortunately for the NCB, it was discovered that the exhaustion of older seams and pits was occurring at a slower rate than had been declared two years earlier, and that there existed many additional opportunities to modernise and reconstruct existing mines. Planned output from existing pits for 1985 was in consequence raised from 63 to 89 million tonnes. The production target for the very oldest and most expensive mines, however, was lowered from 37 to 21 million tonnes. In its subsequent evidence to the Commission on Energy and the Environment in January 1979, the Board confirmed that (even without further major investment decisions) its 1985 deep mine capacity would be 120 million tonnes, and therefore 'on target' (Table 1.1).

Table 1.1 The production plans of the NCB for 1985, 1990 and 2000 (million tonnes).

	1985	1990	2000
output from deep-mine capacity, with no further approvals for further major investments	120	105	80
incremental output available from unapproved projects at existing deep mines	—	9	15
opencast production	15	15	15
Sub-total (without further new mines)	135	129	110
identified new mines, not yet approved	—	16	18
other new mines, not yet identified	—	—	37
Total	135	145	165

Source: NCB evidence to the Commission on Energy and the Environment (1979), except opencast figures which are taken from Tripartite Group (1979).

In that same document, the NCB argued that as a result of exhaustion the industry's underground capacity was likely to fall at a rate of 3 million tonnes per year between 1985 and 1990 to 105 million tonnes, unless they were to make further investments in addition to those already approved. Consequently the Board decided that, in order to meet its targets, a further 9 million tonnes (of as yet unapproved capacity) would be required at existing mines, and an additional 16 million tonnes would have to be provided at 'identified' new mines such as the three for which planning permission had been requested in the Vale of Belvoir, all to come 'on stream' before the end of the 1980s. With opencast mining activity running at 15 million tonnes per year, this would give the country an overall coal production capability of 145 million tonnes in 1990. Looking further ahead in their evidence to the Commission on Energy and the Environment in 1979, the NCB saw existing capacity declining still further through natural exhaustion from 105 million tonnes in 1990 to 80 million tonnes in 2000. The view was taken however, that additional coal produced

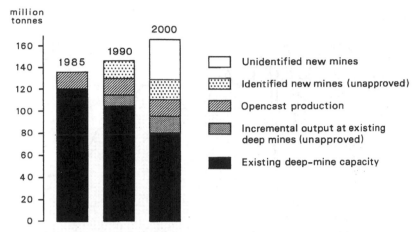

Figure 1.1 The production plans of the NCB for 1985, 1990 and 2000.

from (as yet) unapproved projects at existing mines could add a further 15 million tonnes to this capacity, and that opencast mining would continue to provide *at least* another 15 million tonnes each year. With a production target of 165–70 million tonnes in the year 2000, this left the Board with the need to plan for the development of some 55 million tonnes of new mine capacity to come into production in the 1990s, only 18 million tonnes of which had been identified in 1979 (Table 1.1 and Fig. 1.1).

Given the history of the British coal industry since World War I, a period during which its output has substantially, albeit erratically fallen (Fig. 1.2), and given especially the rapid contraction of the industry in the 1960s when both the number of mines and the workforce were halved and production fell from 184 to 134 million tonnes, this is an exceptionally bold programme of expansion. It will involve the commitment of enormous sums of public capital expenditure. Without quite dramatic improvements in productivity, expansion will demand the recruitment of a very much larger mining workforce than the 230 000 employed in 1980. It will occasion the spread of coal mining activities into localities previously without such a tradition, necessitating the construction of new houses and towns, schools, hospitals, shops, churches, roads and railways, dramatically altering the appearance of some parts of the country. This programme of expansion is based upon the presumption that there will be a marked reversal of the fortunes of the coal industry in the markets for energy, and that British consumers will turn increasingly to the NCB to satisfy their energy needs.

At first sight of course, the plans of the Board are very much in tune

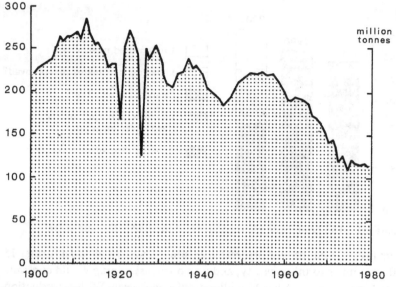

Figure 1.2   The production of coal in Britain, 1900–79.

with the times. In June 1979 the Heads of Government of the major industrialised nations of the West met in Tokyo and agreed that 'the most urgent tasks are to reduce oil consumption and to hasten the development of other energy sources'; their final communiqué also pledged 'to increase as far as possible coal use, production and trade' and to 'maintain positive attitudes towards investment for coal projects.' Earlier, in May 1979, the International Energy Agency (IEA) issued its 'Principles for IEA Action on Coal' and affirmed the importance of solid fuels in the satisfaction of the world's future energy demands: 'beyond 1985 coal could provide a substantially greater contribution to the energy needs of IEA countries. This depends on the adoption by governments now of appropriate coal policies which stimulate capital investment on a scale commensurate with the long-term potential of this energy source.' The present Conservative Government, although they have refused to be drawn into a debate on production targets and plans, have nevertheless opined that 'Coal is our greatest single natural resource . . . (and) Coal's prospects have never been better' (Secretary of State for Energy 1980a). 'The Government's strategy for coal, embodied in the Coal Industry Bill (1980), will give full opportunities for NCB production to expand over the years ahead' (Secretary of State for Energy 1980b).

Yet many aspects of the British coal industry's expansion programme

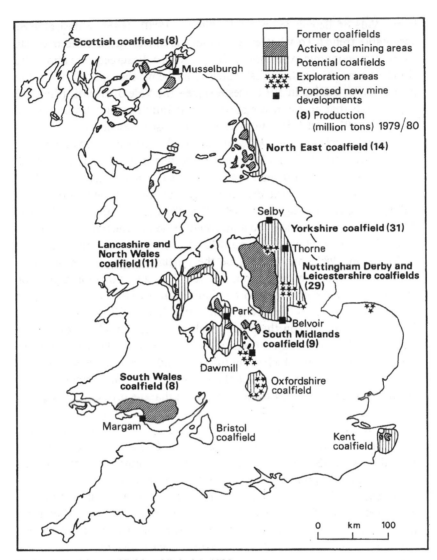

Figure 1.3 The coalfields of Britain, 1980.

are in dispute. There are disagreements, for example, of a technical nature concerning the efficiency and the costs of retreat as opposed to longwall mining methods, and in particular the wisdom of leaving an increasing amount of coal underground after a seam has been mined. With mechanisation, the proportion of coal won from worked reserves has halved in the last 25 years; it currently averages 50%; and the planned recovery rate in the Vale of Belvoir is only 41% (Stocks 1980). Is this a desirable or acceptable trend? To what extent should a conservation ethic overrule short term economic expediency? There are also disagreements about the siting of new mines, the most noteworthy being the opposition to the proposals for mining in NE Leicestershire – the Vale of Belvoir (Fig. 1.3) – proposals that occasioned a major public Inquiry which lasted some 6 months in 1979 and 1980. But perhaps the most important disagreement concerns the scale and the timing of the NCB's modernisation and expansion programme, and the fundamental doubts in the minds of at least some observers that surround the future size of the markets for coal.

Not a few students of the British energy economy believe that the probability of the NCB being able to sell 170 million tonnes of coal in the year 2000 is so remote as to demand a thorough review of the industry's investment plans. Berkovitch (1977, pp. 230–1), for example, notes that 'The period from 1980 onwards will be the crucial time . . . (since) it seems possible that coal may once again find itself in danger of sliding over the hump of the next switchback with a diminishing demand from [its largest customer] the Central Electricity Generating Board (CEGB).' Cook and Surrey (1977, p. 96), on the other hand, are uncertain about the subsequent decade: 'The major decision to be taken in the coal industry concerns the extent to which the industry should plan to increase output in the 1990s and thereafter . . . since there is some risk that demand will not expand rapidly.' In any case it is transparent that the production target of the Board was somewhat arbitrarily determined. It appears to have its origins in an estimate that first appeared in 1977, of the demand for British coal at the end of the century. The Tripartite Group estimate ranged between 135 and 200 million tonnes, 170 million tonnes being the approximate mid-point in that very considerable range.

In 1868 Stanley Jevons, then a fellow of University College London, published his now famous treatise, *The coal question*. His central concern was the speed at which the demand for British coal was increasing, given the finite nature of the energy resource base upon which the domestic coal industry – and indeed the whole of the British economy – was at that time founded. Such a disquiet has no place in coal industry thinking or public discussion today. Jevons' assumption about the exponential growth of demand was ill-founded, and, as has been noted, the production of coal has fallen rather than increased in recent decades. Moreover, in 1977 it was authoritatively claimed by the industry itself that the country's technically recoverable reserves of coal amounted to 45 000 million tonnes, enough to support the existing rate of production for over 300 years

(Tripartite Group 1977). The present coal question, therefore, is not how much coal can be produced, and for how long, but how much can be (profitably) sold over the next 20 or so years. The 1973–4 oil price rise gave the coal industry a renewed and substantial competitive advantage. *Plan for coal* asserted that 'on the basis of commercial pricing the coal industry has now the capability for the first time for many years to bear its full production costs and still compete overall with oil' (NCB 1974). Industrial users of coal paid 9% more per therm than did industrial users of oil in 1973; but in 1974 industrial coal prices came to be approximately half those of oil, and although the gap subsequently narrowed, four years later they were still only about two-thirds those of oil. Nevertheless, British coal consumption in 1979 was 4 million tonnes *below* that of 1973 – 129 million tonnes as compared with 133 million tonnes – and production had *fallen* by nearly 8 million tonnes (from 130 to 122 million tonnes) as net coal exports of 1 million tonnes were replaced by net imports of 2 million tonnes.

It is against this background that the present study explores in some detail the market prospects of the British coal industry over the next two decades. The market tendencies and future probabilities that the study reveals, and the doubts that it raises about the wisdom of the current investment strategy of the NCB, undoubtedly justify serious public and Government consideration. The evidence also poses a major challenge to the management and the unions of the British coal industry. In drawing attention to the most likely levels of domestic and overseas demands for British coal during the next 20 or so years, the study demonstrates that the market opportunities open to the NCB are likely to contrast quite vividly with those assumed in its expansion plans. More specifically, the evidence shows that the demand for British coal at home and overseas will probably stabilise at a lower level than that hitherto indicated by both the Board and the Department of Energy. Moreover, if the demand for coal subsequently expands, it could well do so at a later date than that which underlies current coal industry planning – and might be well into the 21st century. Assuming, therefore, that reasonable success attends those plans already approved by the NCB, and in part already being implemented with the objective of stabilising and then expanding British coal production, and short of the 'premature' closure of many existing mines, the evidence leads to an initial and central conclusion. This is that, without revision, part of the NCB's modernisation and expansion programme is likely to be a huge and highly speculative investment dedicated to the proposition that either the markets for British coal must respond to the production ambitions of the mining industry, or that the economic and social costs of overcapacity plus the 'premature' closure of many older and higher cost mines is of no consequence to present investment decisions. Such a proposition is untenable in economic logic and, without much more deliberate consideration, unacceptable in the national interest. There exist alternative and more acceptable strategies which the coal industry could pursue, strategies which would still allow the industry to modernise and increase its productivity, which would still leave the industry as a major contributor to national energy supplies, and which would still ensure that Britain remained one of the largest producers of hard coal in the world – in 1980 it was the fifth, after the United States, the Soviet Union, China and Poland – and certainly the largest within Western Europe.

The sequence of the argument which follows is; first, to examine the general energy prospects for Britain through to the year 2000, drawing particular attention to the key assumptions that underlie all energy demand forecasts, and second, to explore some of the consistencies, inconsistencies and ambiguities of energy policy statements by governments in recent years. The discussion then turns to the recent performance of the British coal industry in the satisfaction of domestic energy demands, and examines in some detail its market prospects to the turn of the century. The central and continuing importance of the power station market for coal is heavily underlined. Next, and relatively briefly, the probability of net coal imports into this country is considered and the overseas demand for British coal reviewed. Against this background, the plans and the future prospects for coal production in Britain are re-examined. The conclusion of the study is that, on the basis of the best independent information and analyses currently available – and particularly in the light of the most recent projections of the Department of Energy and statements by the Central Electricity Generating Board – for at least a decade and probably longer there is unlikely to be an expanding market for British coal. In the light of experience since 1974, a full awareness of the market and institutional constraints surrounding the British coal mining industry, plus any reasonable set of assumptions about Britain's economic growth prospects and prospective energy requirements, it is concluded that the time is now ripe to reconsider the magnitude and the timing of the NCB's investment programme.

2 *Future national energy requirements*

At the National Energy Conference in 1976, the Department of Energy estimated that the total energy requirements of the United Kingdom in the year 2000 could be as high as 650 million tonnes coal equivalent (m.t.c.e.). Within three years, however, the Department had revised its forecasts downward and, in their *Projections 1979*, postulated a demand for energy at the end of the century within the range 445–510 m.t.c.e. A few months later, at the Vale of Belvoir Inquiry, the Department revealed yet another year 2000 forecast, this time based upon a highly pessimistic assumption about the country's future growth prospects: it postulated a requirement for less than 400 m.t.c.e., which is only 6% more than the 1979 (gross) inland consumption of 372 m.t.c.e.

One of the major effects of the 1973–4 international oil price rise and the re-establishment of those levels (in real dollar terms) in 1979, has been to slow actual growth and to render more uncertain future economic growth throughout the industrialised world. The precise effect of energy price increases since 1973 cannot be accurately quantified with any confidence. The prospective impact of the more recent oil price increases, and possibly even higher real prices in future, therefore, is obscured by many unknown relationships. Just how the major oil importing countries will respond to their altered terms of trade and energy supply instability, and how the oil producing countries will adjust the levels of their non-oil imports, remain very open questions. It is widely expected, however, that as an associated effect (if not as an immediate consequence) of the higher 1979 and 1980 oil prices following the Iranian crisis, the already slowed growth of total output in the industrialised world will be reduced still further in the next few years.

In their comparison of energy projections to 1985, Brodman and Hamilton (1979) have demonstrated the growing international awareness of these altered economic circumstances, and how they have affected 24

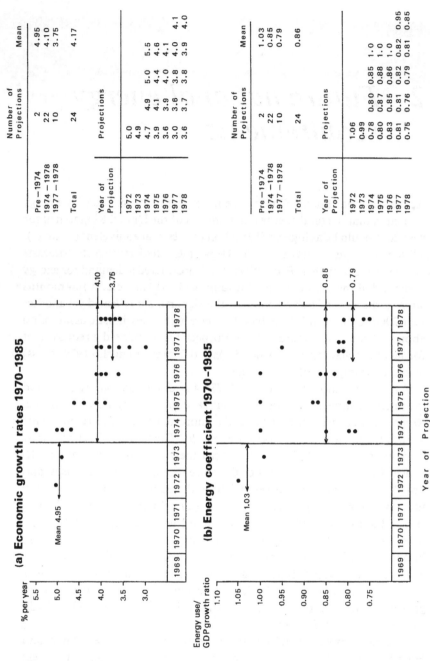

Figure 2.1 Projections of OECD economic growth rates and energy coefficients, 1970–85.

energy demand projections for all or part of the Organisation for Economic Cooperation and Development (OECD) group of countries (Fig. 2.1).
The mean of the Gross National Product or Gross Domestic Product
(GNP/GDP) growth rate assumptions of the 24 projections for the period
1970–85 is 4.17% per annum; however, those made after 1973 assume
substantially lower growth rates (3.9%) compared with those made earlier
(5.0%). The estimates prepared immediately after the oil crisis continued
to reflect the same optimism as those made before the crisis, but in 1975
and 1976 they fell to an average of 4.08%, and in 1977 and 1978 they fell
yet further to 3.75%. Even accepting the uncertain relationship between
higher energy prices and economic growth, it would be reasonable to
assume that more recent projections of economic growth in the OECD to
1985 would in all probability adopt even lower prospective rates of
growth. The growth rate assumptions used in the IEA's *Steam coal prospects to 2000*, which were estimated in 1978 and which show an average
growth of 3.9% per annum for the whole of the OECD between 1976 and
1985, would therefore be regarded as highly optimistic today. The same
would apply to the component national figures (see Table 2.1).

Table 2.1 Actual and forecast growth of GNP in some countries of the OECD
(% per annum).

	Actual	*Forecast*		
	1966/67–1976/77	*1976–85*	*1985–90*	*1990–2000*
Canada	4.7	4.1	3.4	3.1
France	4.6	4.0	4.1	4.1
Germany	3.6	3.7	3.7	2.9
Italy	3.9	3.6	3.8	3.5
Japan	7.8	5.9	5.0	4.0
UK	2.1	2.7	2.1	2.2
USA	2.8	3.7	2.9	2.2
Total OECD	4.2	3.9	3.5	3.0

Sources: Financial Times 20 December 1979; IEA (1978), p. 16.

A related effect of rising world energy prices has been a widespread
search for energy savings, both in the conservation of energy in its use,
and in higher efficiencies in the conversion of one energy source to
another. This is rapidly altering the so-called 'energy coefficient', which is
the relationship between economic growth and energy consumption.

Once again, Brodman and Hamilton (1979) illustrate the widespread change in expectations concerning future energy coefficients as revealed in recent OECD energy demand projections (Fig. 2.1). Two made before 1973 use a figure of 1.03 compared with the historical OECD coefficient of 1.05 recorded during the 15 years prior to 1973. Energy projections made after 1973, by contrast, assume a significantly different relationship, with a mean coefficient of 0.85; and the most recent projections examined (made in 1977 and 1978) use 0.84 and 0.79 respectively. It would be surprising if projections made in 1979 and 1980 have not adopted even lower coefficients. Just how far the coefficient will fall is a matter of considerable uncertainty. For the EEC as a whole, for example, the pre-1973 ratio was about 1.0; between 1973 and 1978, however, it was merely 0.04, a 10.24% increase in GDP being accompanied by a 0.42% increase in energy consumption (Table 2.2). Such a low ratio reflects many once-and-for-all effects of the oil price rise. Looking to the medium term future however, it is not at all surprising that some observers such as White (1978) have suggested that the Western European ratio could fall to within the range 0.7 to 0.5 by the year 2000. The outcome will depend, of course, upon the extent to which energy conservation measures are adopted, and the successes that are achieved in increasing the efficiency of energy conversion processes. Within individual countries, the ratio could fall significantly below these levels.

In the British case, the combined effect of lower economic growth

Table 2.2   The energy coefficient in the EEC, 1973–8.

	A	B	B/A
	Growth in GDP (%)	Growth in energy consumption (%)	Energy coefficient
Belgium	+11.40	−1.67	−0.15
Denmark	+7.57	+3.06	+0.40
France	+9.71	+2.95	+0.30
Germany	+15.03	+0.95	+0.06
Ireland	+18.89	+2.41	+0.13
Italy	+10.49	+2.17	+0.21
Luxembourg	+0.14	−15.60	−111.43
Netherlands	+12.35	−1.97	−0.16
UK	+4.35	−3.23	−0.74
EEC	+10.24	+0.42	+0.04

Source: Commission of the European Communities (1979a).

Table 2.3 Primary fuel demand in the UK in 1979, and alternative forecasts for 1985, 1990 and 2000 (million tonnes coal equivalent).

Date of forecast		Actual 1979	Forecasts					
			1985		1990		2000	
			range	mid-point	range	mid-point	range	mid-point
	Actual	372						
1974	*Plan for coal* (1974)		400–490	445	—	—	—	—
1976	National Energy Conference		350–400	375	—	—	500–650	575
1976	*Energy policy review* (1977)		380–450	403	420–525	473	500–650	575
1977	*Coal for the future* (1977)		—	—	—	—	500–650	575
1977	*Energy policy* (1978)		390–415	402	—	—	450–560	505
1978	CEGB *Corporate plan* (1978)		352–420	386	379–477	428	447–633	540
1978	International Institute for Environment and Development (Leach *et al.* 1979)		—	—	358–380	369	330–361	346
1979	Department of Energy, *Energy projections 1979*		—	—	410–437	424	445–510	478
1979	Department of Energy, *Energy projections 1979* (extended and revised*)		—	—	355–437	396	395–506	451
1979	Robinson (1980)		374–387	381	383–407	395	403–450	427
1979	WOCOL (1980)		350–400	375	375–425	400	400–500	450

* These forecasts take into account the Department of Energy's 1% per annum growth case, and their revised (January 1980) expectations of coking coal demand.

prospects and lower energy coefficients has been reflected in the total primary fuel demand projections and forecasts of the Department of Energy. It has been mirrored even more vividly in the forecasts of certain independent observers (Table 2.3). For example, for the year 1985, *Plan for coal* in 1974 forecast a primary fuel demand ranging between 400 and 490 m.t.c.e., with a mid-point of 445 m.t.c.e.; by 1977, the Department of Energy in *Coal for the future* had lowered the forecast range to 380–450 m.t.c.e., with a mid-point of 403 m.t.c.e.; and more recently the CEGB (1978), Robinson (1980) and the WOCOL (1980) study have suggested even lower ranges with mid-points of 386, 381 and 375 m.t.c.e. respectively. Similarly, for the year 2000, the Department of Energy in 1976 (for the National Energy Conference) forecast a primary fuel demand lying within the very wide range of 500–650 m.t.c.e., with a mid-point of 575 m.t.c.e. By 1979, however, the mid-point of a new Departmental forecast was 478 m.t.c.e. – within the range 445–510 m.t.c.e. This latter figure was based upon a highly optimistic assumption that economic growth in Britain would be sustained at an annual average rate of 2.7% for the rest of this century. The Department also made a projection in 1979 on the basis of a 1% per annum growth rate. Taking this into account, plus the government's (January 1980) revised expectations concerning the British steel industry and its coking coal requirements in the next 20 years, the

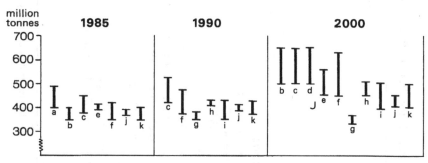

a	Plan for Coal (1974)	g	International Institute for Environment and Development (Leach et al 1979)
b	National Energy Conference (1976)		
c	Energy Policy Review (1977)	h	Department of Energy, Energy Projections (1979)
d	Coal for the Future (1977)	i	Department of Energy, Energy Projections (1979) (extended and revised)
e	Energy Policy (1978)	j	Robinson (1980)
f	CEGB Corporate Plan (1978)	k	WOCOL (1980)

Figure 2.2   Alternative forecasts of primary fuel demand in the UK for 1985, 1990 and 2000.

Department of Energy's forecast range widened to 395–506 m.t.c.e. with a mid-point of 451 m.t.c.e. The WOCOL (1980) figure was approximately the same. The evidence submitted to the Belvoir Inquiry by Robinson (1980), however, proposed an even lower range, with a mid-point of 427 m.t.c.e., and Leach *et al.* (1979) have suggested that it would be possible to contain energy demands within the range 330–361 m.t.c.e. (with a mid-point of 346 m.t.c.e.) and still achieve an acceptable rate of economic growth (see Fig. 2.2).

Table 2.4 A forecast of the markets for coal in the UK in 1985, 1990, 1995 and 2000: the Department of Energy 1978 Reference case (million tonnes).

	Actual	*Forecasts*			
	1975	*1985*	*1990*	*1995*	*2000*
power stations	73.4	88.2	103.4	103.4	64.1
coke ovens	18.6	22.4	23.9	25.4	26.8
manufactured fuel	4.0	1.6	1.1	0.8	0.6
collieries	1.2	0.5	0.4	0.4	0.4
industry	9.6	9.7	12.1	20.8	41.0
domestic	11.5	5.9	4.4	3.2	2.5
other consumers	1.9	1.2	1.1	1.0	1.0
SNG production	0	0	0	0	0
Total	120.2	129.5	146.4	155.0	165.3

Source: Chief Statistician, Department of Energy.

Turning more specifically to coal, the Department of Energy has in recent years tentatively and hesitatingly lowered its expectations of future British demand in line with the shifting prospects for energy overall. For the year 1985, for example, a market for 125–150 million tonnes of coal was proposed by the Board and approved by the Government in the 1974 *Plan for coal*. This range was narrowed to 125–135 million tonnes in the Tripartite Group's 1977 *Coal for future*. The upper of these two figures was adopted as a 'supply' figure in the 1978 Green Paper on *Energy policy*, but the estimated demand for coal in that government statement – the so-called 'reference case' – was only 129.5 million tonnes (Table 2.4). In the central forecast of the Department of Energy's *Energy projections 1979* there were no estimates of coal demand in 1985. However, for the year 1990 coal requirements in Britain were put at 124–132 million tonnes, a range considerably below the 146 million tonnes projected in the reference case of the Green Paper a year earlier. In the Department's extended and revised forecasts (extended to include the 1% growth case, and revised to take note of changed expectations with regard to the steel industry) the range was widened to 105–132 million tonnes, with a mid-point of 119 million tonnes. This figure is slightly *less* than demand in 1978, and along with the other forecasts for 1985 and 1990 is set out in Table 2.5.

Table 2.5   Alternative department of energy forecasts of total coal demand in the UK in 1985, 1990 and 2000 (million tonnes).

Date of forecast		1985		1990		2000	
		range	mid-point	range	mid-point	range	mid-point
1974	Plan for coal (1974)	125–150	138	—	—	—	—
1977	Coal for the future (1977)	125–135	130	—	—	135–200	168
1977	Energy policy (1978)	—	129.5	—			
1979	Energy projections 1979	—	—	124–132	128	128–165	147
1979	Energy projections 1979 (extended and revised)	—	—	105–126	116	110–159	135

Similarly, for the year 2000, the Department has revised its expectations of the coal industry's market prospects downwards. A market for 135–200 million tonnes, with a mid-point of 168 million tonnes, was proposed in 1977 in the Tripartite Group's *Coal for the future*. (This was the range that appears to have been narrowed to a market forecast of 170 million tonnes given in evidence to the Commission on Energy and the Environment by Sir Derek Ezra on behalf of the Coal Board, early in 1979.) Somewhat earlier, however, in the reference case of the 1978 Green Paper, the Department of Energy had estimated year 2000 demand at 165 million tonnes. In their *Energy projections 1979*, the Department published a forecast range of 128–165 million tonnes, with a mid-point of 147 million; but when the 1% growth case was taken into account, along with the most recent interpretation of the steel industry's prospects, the range widened to 110–159 million tonnes, and the mid-point fell to 135 million tonnes – the production target originally espoused by the NCB for 1985. These figures are also summarised in Table 2.5.

Although the Department of Energy has adjusted its projections downwards over the last two or three years, it has to be questioned whether it has been bold enough. The Department continues to assume that the energy coefficient will range between 0.45 and 0.5 in the year 2000, as compared with a pre-1973 level of approximately 0.66, and a post-1973 negative coefficient of −0.7. Yet there are good grounds for agreeing with Robinson in the belief that, prospectively, the coefficient could fall to 0.4 or 0.35 as a result of not only more efficient modes of energy use, but also the changing structure of the British economy. As the manufacturing sector (and particularly the 'heavier' energy intensive manufacturing industries such as steel making, non-ferrous smelting, heavy engineering, ship building and foundry operations) become of decreasing importance to the British economy, and as the 'lighter', high technology manufacturing firms and the service industries expand, so it can be expected that economic growth will be paralleled by ever-smaller increases in the country's energy requirements. This is particularly likely if, as is widely expected, rising real energy prices encourage substantial improvements in the care and efficiency with which energy is used. While the future energy coefficient in Britain cannot be forecast with any significant degree of confidence, lower coefficients than those currently used by the Department of Energy need to be examined in any comprehensive set of energy demand forecasts.

It is important to question how far the NCB have taken note of the changed expectations, nationally and internationally, in regard to both economic growth and energy demand. It is particularly significant that the Board failed to provide the Vale of Belvoir Inquiry with original and convincing evidence to demonstrate their full awareness of the substantial changes in the economic and energy environment since their expansion plans were first forged. A reconciliation of the 1974 and 1977 ambitions of the Board with a contemporary reading of future energy requirements is far from easy, yet the aim of the NCB to produce and sell 135 million tonnes of coal in 1985 and 170 million tonnes in the year 2000 appears to remain unchanged. Only with persuasive evidence that British coal now has the prospect of becoming more competitive in the national energy market than was thought likely to be the case a few years ago could it be demonstrated that this apparent inflexibility is rooted in sound market judgements. Again, no such evidence has been produced for public scrutiny by either the NCB or its supporters. Implicit in the thinking of the Board appears to be a notion that British coal is likely to capture an increased share of the substantially reduced expectations of energy demand growth in the next 20 years. Such a proposition cannot be left unchallenged.

Indeed, it is important to remember that the higher price of oil since 1973, and the associated disequilibria in oil markets, in Western Europe at least, has not automatically occasioned an increase in the demand for coal in absolute terms – especially for domestically produced coal. In the countries of the EEC between 1973 and 1978, for example, the demand for coal actually fell from just over 303 million tonnes to 287 million tonnes, and a diminishing share of this demand was supplied from the mines of the Community itself as net imports rose by 50% from 30 to 45 million tonnes. This decline in coal requirements reflected in part the immediate fall and the slow recovery of total energy consumption after the oil crisis and the subsequent recession. However, it is noteworthy that during the same period (1973–8) EEC consumption of natural gas rose 57% from 162 to 225 m.t.c.e., and the expansion of nuclear power was more than 100% as output grew from 19 to 41 m.t.c.e. Similarly, in Britain, coal consumption fell from 131 to 120 million tonnes between 1973 and 1978, whilst the use of natural gas increased nearly 50% from 44 to 65 m.t.c.e., and nuclear power contributed an extra 3 m.t.c.e. to the energy supply (making its contribution 13 m.t.c.e. in the latter year).

It has to be asked, therefore, whether the NCB's expectations of future demand for their coal, and even perhaps the Department of Energy's analyses, too readily embrace an undeclared assumption that the growing scarcity and the rising price of oil will *automatically* generate new demands for solid fuels. That assumption needs to be rigorously examined in the light of recent market experience and careful market analysis. Before doing so, however, it is important to clarify the nature and the ambitions of public policies in the energy sector. After all, on 25 February 1980 a Minister of State in the Department of Energy spoke of the need to create 'a modern, efficient, low cost, highly productive (coal) industry capable of meeting expanding demand in the future. The Government are committed to the high level of investment needed to ensure this' (*Hansard*, para 1023). Subsequently, during the second reading of the 1980 Coal Industry Bill, the Secretary of State for Energy

(1980a) spoke of the industry's 'opportunity to secure for itself a prosperous and good future (based) on new market opportunities and efficient and competitive production'.

3 Energy policies

Energy policy statements, certainly those by national governments rather than by international bodies, tend to comprise three distinct elements. First, they include some assessment of future demand and supply probabilities through the use of projections, estimates and forecasts. Second, they present a summary of government attitudes towards those probabilities, a statement of public goals in regard to energy markets and industries, plus a declaration of a preferred outcome of events. And, third, they announce a series of measures that, it is hoped, will help to bring about that preferred outcome. It has to be said immediately, of course, that the second element in policy statements is not always followed by a sufficiency of the third! Such policy statements seek to ensure a consistency of attitudes and actions by government towards different energy sources; they provide a framework for taking decisions about government-owned or government-regulated energy industries; they seek to anticipate, alleviate and possibly overcome problems in the energy sector; and they invariably hope to modify and sometimes improve the workings of the market.

The 1978 Green Paper, *Energy policy – a consultative document*, set out the interim views of the Labour Government towards energy futures in general and the coal industry's prospects in particular. It was subsequently supplemented by a set of 1979 projections and commentary by the Department of Energy, and by a succession of statements by Secretaries and Ministers of State in both the Labour and subsequent Conservative Governments. That the coal industry should seek to embrace, as quickly as possible, higher productivities and lower costs through investment in more modern facilities and technologies is widely agreed. However, it is also the case that by the end of the 1979–80 parliamentary session, the Conservative Government, although regularly endorsing the notion of further investment in coal industry modernisation, had not declared any policy objective with regard to the scale and the timing of the industry's expansion. Even after all the stages of the Coal Industry Bill had been passed, a full statement comparable to that made on nuclear power on 18 December 1979 (Secretary of State for Energy 1979) had not been made. Only in a guarded parliamentary written answer (*Hansard*, 22 July, Col. 139) in which reference was made to 'Current projections of coal production in the longer term are between 127 and 138 million tonnes in 1990, and between 137 and 155 million tonnes in 2000' has the Government been willing to imply something of its thinking about the scale of the industry in the medium term.

The objectives of energy policy, of course, tend not to alter dramatically from government to government. In the past, all governments have sought to ensure adequate energy supplies; to hope that these would be efficiently used; and to see that these two goals would be achieved at the lowest practicable cost to the nation. At the same time, successive government policies have confirmed the importance of consumer sovereignty: in the Green Paper it was reasserted that 'freedom of the consumer to choose between fuels . . . should, where possible, be maintained and increased'. Policies have also repeatedly underlined the qualified importance of self-sufficiency:

> Self sufficiency . . . is desirable only insofar as indigenous sources may offer supplies which have a lower resource cost, are more secure, or both, as compared with imports; it is not an objective in its own right. (Secretary of State for Energy 1978, p. 4).

In seeking to explain energy policies in recent years, the Department of Energy and government spokesmen have referred repeatedly to three 'main components' of policy, one of which is the encouragement of conservation. In the Green Paper on *Energy policy* in 1978, for example, energy savings from conservation were assumed likely to be some 20% of demand by the year 2000. In the more recent *Energy paper no. 33* of the Department of Energy (DEn 1979b), it was reaffirmed that 'Conservation is one of the three main components . . . of the Government's energy policy in the long term'. A 20% saving was also incorporated in the *Energy projections 1979* of the Department of Energy. In fact, considerable energy savings have already been made since the 1973–4 energy crisis. It has been estimated that, after allowing for changes in economic activity and the weather, savings in primary energy use in 1975 may have been in the region of 6% of the amount that would probably have been consumed in that year without conservation – though the exact figure was uncertain and could have been as low as 2% or as high as 10% (DEn

1976). In the Green Paper on *Energy policy*, the Department of Energy stated its belief that approximately the same level of saving had been achieved in subsequent years, and that 'with a really vigorous and sustained Government programme still larger savings would be achievable'. The potential savings are undoubtedly considerable, and they have been well illustrated by the work of the International Institute for Environment and Development. In *A low energy strategy for the United Kingdom* (Leach *et al*. 1979), it is proposed that quite remarkable energy savings are within the country's technological grasp, and that total energy use in Britain in the year 2000 need be no higher than it is today. Even with a truncated nuclear power station programme, it is calculated that 'Coal production need be only some 120 million tonnes a year (in 2000)', and that only a small additional tonnage would be required in the subsequent 25 years.

Conservation, substantially aided by energy price rises in real terms, is undoubtedly the least contentious aspect of government energy policies – and is likely to remain so. It is reasonable to assume, therefore, that some of these very considerable additional energy savings will in fact be achieved, and that total energy requirements in Britain could well be reduced to a level below that currently assumed in government projections of demand. Whilst the major conservation effort will obviously be directed towards the consumption of oil, there can be little doubt that the same policies will also affect the demand for electricity and coal.

The two other 'main components' of energy policy, as declared in *Energy paper no. 33*, are coal and nuclear power (Department of Energy 1979b), and they were readily embraced by the incoming Conservative Government. In the Second Reading of the Coal Industry Bill, the Secretary of State for Energy stated that the Government 'have put coal, along with higher energy efficiency through conservation, along with nuclear energy, at the centre of our energy policy' (*Hansard*, Col. 1377). In one sense coal and nuclear power play *complementary* roles; both will be required in some considerable measure to satisfy future national energy needs. Each clearly has a distinctive and preferred place in the production systems of the CEGB and the South of Scotland Electricity Board (SSEB). At the same time, however, it cannot be overlooked that coal and nuclear power are also in *competition* to satisfy certain parts of the future base and middle loads of the electricity generating Boards. The importance of this competition is noted in Chapter 4. At this stage it is sufficient to note that, whilst government policy statements confirm the future importance of both these sources of energy, choices will have to be made between them at different times and in the context of particular markets. These choices will be influenced above all by the opinions and preferences of the electricity generating authorities, who in recent years have shown a clear desire for more nuclear power rather than coal. It follows, therefore, that broad Departmental and government statements such as 'Even allowing for a maximum future contribution from nuclear power, there will be a large and continuing need for coal' (DEn 1979a) in no way guarantee either a particular magnitude or even an upward trend in the aggregate demand for coal in Britain. Indeed, it is important to note how highly qualified recent statements by the Department have become on the question of future coal markets. That same document suggests that 'Demand for coal during the 1990s *is likely to be* at least at present levels and *the chances are* that the need to use and produce coal will be rapidly rising by the end of the century' (author's italics).

Table 3.1 Expected and planned capacity of nuclear power stations in Britain, 1980 and 2000 (MW).

	1980	2000
magnox stations		nil
AGR under construction 1980 (Dungeness B, Hartlepool, Heysham, Hinkley Point, Hunterston B)	} 6300	} 6250
AGR with investment approval (Heysham B, Torness)	—	2500
new programme announced 18 December 1979	—	15 000
nominal capacity	6300	23 750
probable capacity*	n.a.	22 500

* Technical difficulties with the first AGR programme demand some reduction in their estimated capacity in 2000.
Source: CEGB (1980d).

By December 1979 the new Conservative Government had made a detailed statement on its nuclear power plans. In addition to the five Advanced Gas-Cooled Reactors (AGR) under construction, and the two further AGRs given investment approval by the previous Labour Government, the Secretary of State announced that there would be a further 15 000 MW of capacity built on the basis of one new station being ordered each year from 1982. In the absence of further orders, and following the retirement of the first generation Magnox reactors, this programme could result in nuclear plants having a nominal capacity of nearly 24 000 MW (Table 3.1). Given the technical difficulties of the first AGR programme, however, it is prudent to assume that the effective capacity of British nuclear stations in the late 1990s would be 1250 MW lower.

Whilst the Department of Energy in its publications, and government spokesmen in their statements, have tended to single out three 'major components' of energy policy – conservation, coal and nuclear power – it is imperative not to overlook the major importance and the prospectively greater role of natural gas in British energy supplies. This source met 20% of the country's energy needs in 1979 (71 m.t.c.e.), and seems destined to play an increasingly important role in the British energy economy as additional reserves are made available from under the North Sea. Supplies principally from the southern sector fields and from the Frigg and Brent reserves to the north will reach an equivalent of 75–80 million tonnes of coal (over 6000 million cubic feet per day (m.c.f.d.)) in the early to middle 1980s (Fig. 3.1). This figure was adopted as a central estimate in the Green Paper, where an even higher level of supply was also contemplated on the basis of rapid depletion. The Department of Energy built into its 1979 projections the assumption that natural gas availability will tend to fall away in the 1990s. Yet firm grounds exist for believing that supplies can, at the very least, be maintained at the level of 6000 m.c.f.d. throughout that decade, and possibly even increased above this figure. The reasons for so thinking are several. First, gas reserves in some considerable measure are a function of price; the price of both domestic and industrial gas is being raised or is about to be raised substantially as a result of government policy; this will inevitably lead to a reassessment of marginal fields, a more intensive exploration effort, and an upward revision of reserves. Second, higher gas prices in Britain will increase the probability of the British Gas Corporation being able to purchase larger supplies of Norwegian gas for which this country is in competition with continental purchasers. Third, since the Green Paper, the government has given firm backing to the gas gathering pipeline which will link not only reservoirs as far north as Magnus and as far south as Josephine, but could also be extended eastwards to serve Statfjord and other Norwegian sources of supply. The trunk pipeline to the Scottish landfall will be some 400 miles long, and although initially designed to handle about 1000 m.c.f.d., it could eventually transport twice that volume of gas. Hopes have been expressed that the pipeline will be commissioned by 1985.

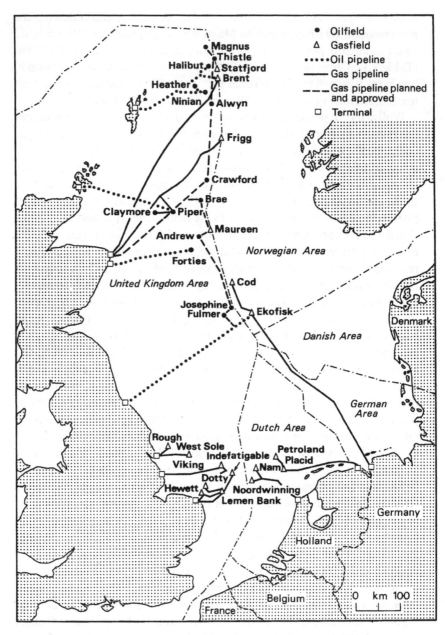

Figure 3.1   North Sea oil and natural gas fields and facilities, 1980.

It is generally agreed that small increments to reserves will be used to extend the time over which 6000 m.c.f.d. of gas is available. More substantial finds, however, in either British or Norwegian waters (or onshore), could see natural gas supplies rising to very much higher levels. At the press conference to present the 1978—9 Annual Report of the British Gas Corporation, the Chairman said:

'Although we remain confident about the prospects for further significant finds of gas around our coasts, we can only rely on the known and contracted reserves . . . these natural gas supplies should be sufficient to meet requirements until the end of the century, or even longer, on the basis of present marketing and depletion policies . . . If, however, major new reserves were to be discovered in the North Sea, or elsewhere around our coasts, the situation could be different, and further expansion above the 6000 m.c.f.d. might be justified.'

It should also be added that such a development would be widely welcomed, even though it would be very much at odds with most government (published) forecasts and pose additional problems for the British coal industry.

This is not the place to enter into a debate about the possible size of British and Norwegian gas reserves. It is nevertheless important to note that, because of its *national* accounting framework, and inherent conservatism, the Department of Energy persistently appears to underdeclare in its Energy Balance Tables the quantities of natural gas likely to be available to British consumers in the future, by ignoring committed and prospective imports from the Norwegian sector of the North Sea. Thus, in the Green Paper on *Energy policy*, 65 m.t.c.e. was noted as the prospective level of (domestic) gas supplies in 1985. Such a figure considerably understates the likely impact of this premium source of energy on British energy markets over the next five years. Supplies of some 85 m.t.c.e. are a much more likely prospect when imports are openly recognised. Similarly, the supply ranges of 68—71 m.t.c.e. in 1990, and 62—5 m.t.c.e. in 2000, that appear in the Department's *Energy projections 1979* not only relate solely to domestic sources of gas, but also gloss over the certainty (and the unchallenged legitimacy) of contracted and additional imports from accessible Norwegian reserves; moreover, they are clearly in conflict with the expectation of the British Gas Corporation that supplies can be maintained at a stable 6000 m.c.f.d. level throughout the 1990s.

Numbers apart, it cannot be disputed that the increasing scale of gas supplies in the 1980s will provide a highly competitive element in the country's industrial, commercial and domestic energy markets – even if, as seems certain, they are priced relatively higher than they are today. In sales to industry, gas prices have moved from an index of 100 in 1970 to 65 in 1973 to 287 in 1979; coal prices, meanwhile, moved from 100 to 132 to 414, and heavy fuel oil prices from 100 to 139 to 710. In real terms, gas prices in 1979 were lower than in 1970 (an index of 84), whilst coal prices had increased to 121 and heavy fuel oil prices to 208 (see Fig. 3.2). The

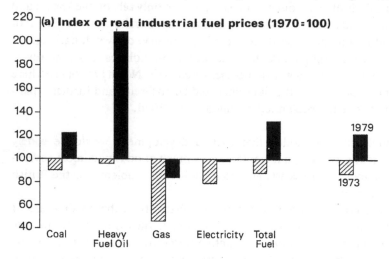

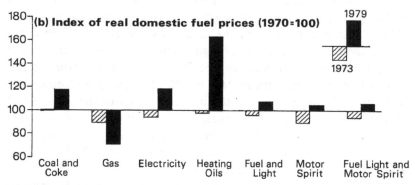

Figure 3.2   Relative energy prices in Britain, 1973 and 1979.

contrasts are equally marked in the domestic sector where the index of real gas prices was as low as 70 in 1979 (1970 = 100), whereas coal prices were 117 and heating oil prices 162 (DEn 1980a). In consequence, substantial increases in gas prices – which will assist in the capture of additional gas supplies – are likely to leave the industry still very competitive against alternative fuels, especially coal. These additional supplies, together with persistent conservation policies, will be employed with growing efficiency. Therefore, although natural gas does not apparently merit a place amongst what the Department of Energy and government spokesmen have called the 'main components' of energy policy, it is very important to note its widening role within the British energy market. It is equally crucial to recognise the considerable constraints that natural gas imposes upon the opportunities available to the coal industry to retain – let alone expand – its markets, especially those in the industrial sector.

Turning to successive policy statements about the future role of coal in the national energy economy, and especially to those in the recent *Energy projections 1979* of the Department of Energy, four points must be made. First, it has to be recognised that the projections, whilst they have been informed to some extent by past and current policy statements (such as that regarding nuclear targets), are *not* in themselves policy; they are not even 'approved and adopted' by the Government; rather they are a framework against which policies and measures might be judged. Second, it is not possible to escape from the fact that coal is very frequently treated in the analyses of the Department of Energy as a 'residual' source of supply, a fuel available to meet energy requirements over and above those not satisfied by alternative fuels. The Department's thinking in its *Energy projections 1979*, for example, is clearly based upon a set of assumptions about future oil and gas supplies, 'a maximum future contribution from nuclear power', and a set of energy demand projections. Together, these assumptions and expectations leave what is seen as 'a large and continuing need for coal'. However, it would appear to be the case that the Department of Energy has not researched the ability or the willingness of energy consumers to accept or use coal, and thereby 'fill the gap' between oil, gas, nuclear and renewable energy supplies on the one hand, and the total projected demand for energy on the other. Its judgements and expectations about the future level of coal sales, therefore, must be regarded as particularly speculative. Third, since the preparation of the 1979 projections, it is clear that several developments in the field of energy have shifted the balance of future probabilities against higher levels of coal demand. There was, for example, the 1979 rise in oil and hence all energy prices; this has helped to depress world economic activity in general (the British economy included) and energy demand in particular. To be sure, 1979 saw coal's price advantage over oil increased once again in many energy markets, and coal sales to the electricity supply industry moved strongly forward. However, the sluggish nature of the rest of the economy reduced the overall level of capital investment (including investment in new energy-using equipment) and slowed the process of substitution from oil to alternative fuels. Fourth, there has been a change in Government and a shift in attitudes and policies towards the exploitation of North Sea oil and gas, which might well have removed some inhibiting factors to further exploration and development. The commitment to coal may not be less, but the desire to exploit more rapidly the assets of the North Sea basin cannot leave the prospects for coal unaffected; in addition, there has been a positive government statement, as has been noted, about the future of nuclear power. The implications of these observations are noted later in the analysis of the coal industry's market prospects.

Turning briefly to the international dimensions of energy policy, various statements have been made by the Commission of the EEC (1980) and a coal policy has been agreed by the member governments of the IEA. All seek to encourage a greater use of coal throughout the EEC and the OECD in the wake of rising oil prices and prospective (relative) oil scarcities. Unaccompanied by agreed measures and appropriate financial support, however, such statements and agreements can do very little except draw attention to changed market circumstances, and serve little to accelerate new investment in energy producing or energy consuming ventures. Certainly within the context of the EEC, it is transparent that the so-called policy of the Commission represents essentially a set of laudable aspirations rather than a likely future reality. This situation stems from the very weak powers of implementation available to the Commission, the unwillingness of the Council of Ministers to vote many of the necessary funds, and the limited convergence of energy policies adopted by the member states. Some loans and grants have been made available, it is true, to facilitate and accelerate the coal industry's modernisation programme: between 1974 and 1979, for example, the NCB attracted over £300 million in low interest loans from the European Coal and Steel Community. However, all the proposals made by the Commission in an effort to increase Western European markets for coal, and the proposed subsidy on steam coal trade within the EEC, have failed to win the endorsement of the Council of Ministers. Community policy, in other words, has failed to reduce the fundamental problem of coal's poor marketability in the EEC. The increased demand for coal that is in prospect is more likely to be satisfied by imports from non-Community countries than by the products of the relatively high cost indigenous industry. Above all, it is clear that the British coal industry cannot look to the rest of the EEC for sales of any magnitude (a point that is explored in greater detail in Chapter 6).

4 *The future power station market for coal in Britain*

In recent years the markets for coal in Britain have shrunk drastically (Fig. 4.1). From sales of nearly 200 million tonnes in 1960, demand at home had fallen to less than 130 million tonnes by 1979. Some markets, such as those of the gas and railway industries, have been totally lost. Other markets – industry and the domestic consumer, for example – have steadily contracted and, in view of the prospective availability of more

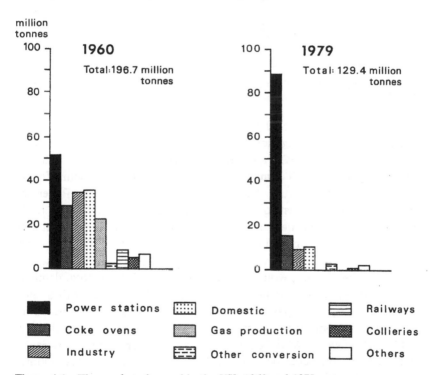

Figure 4.1 The markets for coal in the UK, 1960 and 1979.

natural gas, plus its advantages of cost and convenience, are unlikely to expand significantly in the next decade. The demand for coking coal contracted more slowly at first, but in the last few years it has fallen to *circa* 15 million tonnes. Only the market for power station coal has shown strength and, with some government assistance, has grown from 52 million tonnes in 1960 to 89 million tonnes in 1979 (Table 4.1). It is against this historic experience that future prospects must be realistically judged.

Table 4.1   The markets for coal in the UK, 1960, 1978 and 1979 (million tonnes).

	1960	*1978*	*1979*
power stations	51.9	80.6	88.8
coke ovens	28.2	14.9	15.1
industry	34.9	8.6	9.2
domestic	35.5	10.2	10.5
gas production	22.6	—	—
other conversion	2.3	3.1	2.9
railways	8.9	—	—
collieries	5.0	1.0	0.8
others	6.8	2.0	2.0
Total	196.7	120.5	129.4

Source: Department of Energy, *Digest of UK energy statistics* (annually). London: HMSO.

The crucial market for the coal industry in Britain is that provided by the electricity generating industry. By 1979 nearly 70% of all coal sales in Britain were to the CEGB and the SSEB. This is the market which, it is universally agreed, will dominate the activities of the Coal Board for the rest of the present century. Yet, in preparing the 1978 Green Paper and making detailed forecasts of the evolution of British coal demand to the year 2000, the Department of Energy saw the possibility of power station sales falling back from 89 million tonnes in 1979 to 64 million tonnes at the turn of the century. Under this 'reference case' (Table 2.4), the coal burn in power stations would then represent only 39% of total demand. Despite somewhat less optimistic assumptions about the growth of electricity demand, the Department's *Energy projections 1979* suggested the possibility of somewhat higher power station requirements in the year 2000. The range was 66 to 78 million tonnes, and represented between 47 and 52% of total coal demand (Table 4.2). However, a few months earlier, in its evidence to the Commission on Energy and the Environment, the NCB proposed a market for 90 million tonnes of coal in the country's power stations at the end of the century, a figure equivalent to 53% of their somewhat higher estimate of total coal demand. The considerable divergence between these three forecasts (and others which will be noted later) confirms the need to explore the uncertainties surrounding the future level of coal demand in Britain by the electricity generating industry in very much more detail.

The quantity of coal burnt by the generating authorities in any one year is a function of several factors – total electricity demand, the output of nuclear power plants (whose operating costs are lower), the capacity of coal fired stations compared with alternative fossil fuel generating sets, the relative age and efficiency of the various production facilities, the prices of the alternative primary fuels delivered to the power stations, and government subsidies or aids. During the last few years, although there has been very little growth in the overall demand for electrical energy (sales of 225 267 GWh in 1973, 240 768 GWh in 1979, an increase of less than 7% in 6 years), there has been a significant increase in the amount of coal burnt by the generating authorities – from 76 million tonnes in 1973 to 89 million tonnes in 1979. This increase in coal sales followed the sudden price advantage of coal over oil after 1973–4, and a subsequent period of fluctuating then narrowing price differentials leaving (for the CEGB) a coal : oil price ratio of 0.88 in the financial year 1978–9 (Table 4.3). Coal's advantage widened again with the Iranian crisis in 1980, the coal : oil price ratio standing at 0.58 in January of that year. Three months later, however, following a further increase in coal prices it had narrowed to 0.70 once again. The increasing in coal sales after 1973 was further aided by a succession of government initiatives and subsidies to assist the generating boards in increase their coal burn. Notable amongst these was a coal stocking subsidy, the modification of the (coal and oil) dualfired power station at Kingsnorth to burn more coal, and the planning agreements to use more coal in Scotland and Wales.

This trend of an increasing power station coal burn could, however, be checked in the next year or so and reversed by the middle 1980s – for two main reasons. First, it is by no means certain that coal will be able to retain its current and substantial price advantage over oil. Between 1974 and 1979, the difference between delivered coal and oil prices to the generating boards narrowed to the point where, but for the coal industry's annual Exchequer grants, it might well have disappeared. In an exchange of letters between the CEGB and the NCB in January 1978, the Chairman of the former corporation asserted that, if the price of coal was increased

Table 4.2 Electricity coal burn in Britain in 1978 and 1979, and alternative forecasts for 1985, 1990, 1995 and 2000 (million tonnes).

Date of forecast	1978	1979	1985 range	1985 mid-point	1990 range	1990 mid-point	1995 range	1995 mid-point	2000 range	2000 mid-point
Actual	81	89								
1978 1 Department of Energy, evidence to IEA (1978)			—	73	—	—	—	—	—	—
1978 2 Department of Energy, *Energy policy* (1978), Reference Case			—	88	—	103	—	103	—	64
1979 3 Department of Energy, *Projections 1979*			—	—	89–94	92	—	—	66–78	72
1979 4 Department of Energy, *Projections 1979*, incl. 1% growth case			—	—	80–94	87	—	—	65–78	72
1977 5 CEGB *Corporate plan* (1978)			72–82	77	—	—	—	—	—	—
1978 6 CEGB (August 1979)			62–86	74	71–91	81	62–65	64	35–52	44

Year	Source								
1978	7 CEGB evidence to House of Lords (1978)	62–92	77						
1979	8 CEGB (8 February 1980)	73–89	81	87–92	90	—	—	55–111	78
1980	9 CEGB (28 February 1980)	—	79	79–84	82	—	—	50– 84	67
1977	10 NCB et al. (Tripartite Group 1977)	—	—	—	—	—	—	75– 95	85
1979	11 NCB (1979)	—	—	—	—	—	—	—	90
1979	12 Robinson (1980)	67–71	69	68–73	71	—	—	45–65	55
1979	13 WOCOL (1980)	84–89	87	85–90	87	—	—	86–94	90
1980	14 This study	—	—	75–85	80	—	—	65–75	70
	Assumed annual coal burn of the SSEB	}	7	}	5	}	5	}	5

Table 4.3   Coal : oil price ratios for electricity generation in England and Wales.

(a) Past experience		(b) Forecasts	1985	1990	1995	2000
1970–1	0.93	CEGB *Corporate plan*,	0.94	0.87	0.78	0.71
1971–2	1.07	1978				
1972–3	1.13					
1973–4	0.76	Electricity Council *Medium*	0.91	0.89	0.87	0.84
1974–5	0.65	*term development plan*,				
1975–6	0.77	1979				
1976–7	0.75	CEGB evidence to the	0.72–	0.65–	0.65	0.59
1977–8	0.75	Vale of Belvoir	0.89	0.79		
1978–9	0.88	Inquiry, 1980				
Sept. 1979	0.75	Department of Energy	0.60	0.50	0.45	0.40
Jan. 1980	0.58	evidence to the Vale of				
March 1980	0.70	Belvoir Inquiry, 1980				

Sources: CEGB (1978, 1980d, e); Electricity Council (1979); Department of Energy (1980e).

by much more than 10%, it would cease to have a competitive edge over oil. The SSEB also experienced coal price increases in the 1970s from £6.33 per tonne in 1973–4 to £22.45 per tonne in 1978–9, a rise of 255%; delivered oil prices in contrast rose only 224%, from £15.15 per tonne to £49.15 per tonne in the same period. The future relationship between oil and coal prices is, of course, highly uncertain. Central to this uncertainty is the speed with which oil prices in real terms will increase, and alongside this must be set the question mark overhanging the future costs and prices of coal.

Certainly it can be argued that the rising international worth of Sterling, the rising real costs of mine labour, and the limited success of the NCB's productivity schemes in recent years all suggest that the gap could well narrow again. The Electricity Council (which represents the whole of the electricity supply industry in England and Wales – the Area Boards plus the CEGB) in their *Medium term development plan 1979–86* published in July 1979, took the view that in the medium term coal prices will rise as fast as oil prices. They express the opinion that:

Experience since 1974 has demonstrated that when oil prices rise, the 'headroom' that this creates for coal prices is soon taken up by increases in the costs of coal. If, subsequently, oil prices ease, the increases

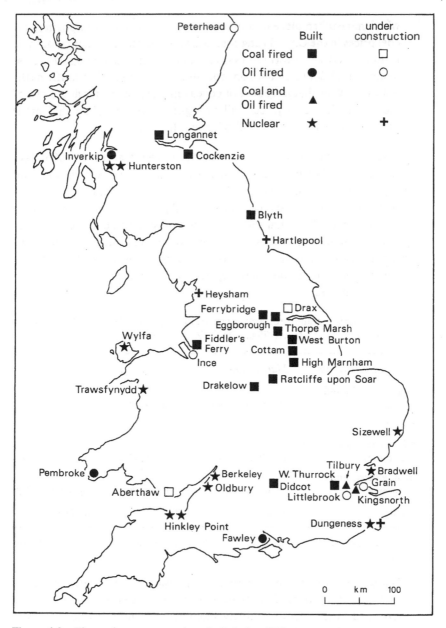

Figure 4.2 The major power stations in Britain, 1980.

which have taken place in coal production costs make it impossible for coal prices in the United Kingdom also to ease, with the result that there follows a series of short term Government measures to support the use of coal which continue until once again oil prices rise. This confirms the industry's view that it would not be reasonable to expect that the rate of increase in UK coal prices will be significantly less than that made possible by increases in world oil prices (p. 21).

On the other hand, both the Department of Energy and the CEGB gave evidence to the Belvoir Inquiry suggesting that the gap between oil and coal prices will steadily widen. The Department, for example, took the view that by the year 2000 the coal *cost* : oil *price* ratio would be as wide as 0.40 (DEn 1970b). That the electricity industry could so dramatically alter its opinion within the space of two years underlines the enormous uncertainties surrounding these matters, and the necessity to explore a range of assumptions about any key determining variable when forecasting energy futures.

Table 4.4   Existing and prospective generating capacity of the public electricity supply industry in Britain, 1979 and mid-1980s (MWe).

Mode of generation	March 1979	Under construction	Total
coal, coal/oil, and coal/gas	47 552 (61.8%)	1 980 (11.3%)	49 532 (56.7%)
nuclear	6 354 (9.1%)	5 280 (30.2%)	11 625 (13.3%)
oil, oil/gas, gas turbines and diesel	13 556 (19.4%)	8 722 (49.9%)	22 278 (25.5%)
hydroelectric and pumped	2 349 (3.4%)	1 500 (8.6%)	3 849 (4.4%)
Total	69 802	17 482	87 284

Sources: Annual Reports and Accounts of the CEGB; SSEB; North of Scotland Hydro Board.

The second reason why the recent expansion of coal sales to power stations might be checked in the near future are the effects, shortly to be felt, of investment decisions mostly taken by the Generating Boards in the late 1960s and early 1970s when both oil and nuclear power appeared to offer considerable and long-term cost advantages over coal for electricity generation. By *circa* 1987 these could add some 17 500 MW of new generating capacity to the public supply system (Table 4.4). Over 30% of this new capacity is nuclear. Nearly half is oil fired, with the largest sets either designed to use fuel oil from adjacent refineries, or, in one instance, located near to the landfall of the oil pipeline from the Brent Field (Figs 3.1 and 4.2). The percentage for oil could perhaps be slightly smaller following construction delays at the Isle of Grain power station and a decision in 1980 to 'mothball' a 1300 MW part of the 3300 MW plant. However, the important point to note is that only 11% of public generating plants now under construction are to be coal fired; this comprises the three 660 MW additional units at Drax in Yorkshire, which at some public expense were brought forward prematurely into the CEGB's construction programme largely to assist with the maintenance of a continuing workload for the power station equipment manufacturers. In terms of its output capacity therefore, the public electricity supply in Britain will significantly increase its absolute and relative dependence upon oil fired and nuclear plant between 1979 and 1987 (Fig. 4.3). The use of oil in

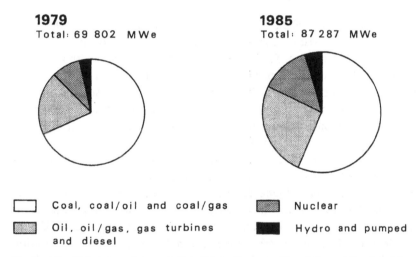

1979
Total: 69 802 MWe

1985
Total: 87 287 MWe

☐ Coal, coal/oil and coal/gas

▦ Oil, oil/gas, gas turbines and diesel

▦ Nuclear

■ Hydro and pumped

Figure 4.3 Existing and prospective generating capacity of the public electricity supply industry in Britain, 1979 and the mid-1980s.

power stations seems almost certain to increase. Meanwhile, the relative importance of coal fired plants will fall from 68% of generating capacity to perhaps 57%. This latter percentage could, in fact, substantially overstate the importance of coal fired plants in electricity supply in the mid-1980s. The nuclear power stations, because of their low operating costs, will serve a major part of the base load, leaving coal (and oil) to meet middle and peak load demands. Further, the next few years will see the closure of some 5 GW of older generating plants with high operating costs, most of which is coal fired; and this will lead to an absolute fall in total coal based generating capacity. In its *Medium term development plan 1979−86*, the Electricity Council foresaw a reduction in coal (and dual coal/oil, coal/gas) generating capacity from 40.4 GW to 38.2 GW between end-1977 and end-1985 in England and Wales.

Firm and proposed plans by the electricity industry to build additional generating capacity for commissioning in the late 1980s point in the same direction. There are no firm proposals for additional coal fired plants anywhere in Britain − although financial provision has been made by the CEGB for the 'replanting', or modernisation, of two existing (but geographically unspecified) coal fired power stations (Electricity Council 1979). As has been noted however, additional nuclear power stations have been approved at Heysham (2 × 625 MW) and Torness (2 × 660 MW); and a further programme of nuclear station construction from 1982 has been announced.

The new nuclear plants, as they are commissioned in the 1980s, will displace all the existing conventional power stations downwards in the generation merit order. The new oil fired power stations, with their high thermal efficiencies, will also displace many of the smaller and older coal fired units of the Boards − unless there is either a very substantial (30% or so) price advantage of coal over oil, or a government decision to maximise the power station coal burn regardless of cost. In 1974−5, for example, when coal prices were particularly advantageous (and there was an average coal : oil price ratio of 0.65), out of 146 coal and oil fired stations in the CEGB system, only seven coal stations were placed higher in the merit order and had a higher load factor than the most efficient oil fired station (Fawley). Thus, it is difficult to see how, following their completion, the new oil fired power stations at the Isle of Grain, Littlebrook, Ince and Peterhead will not make a substantial contribution to British electricity supply. This is particularly the case since these large plants cannot be so readily brought in and out of production, and as a consequence have a 'technical minimum' load factor as high as 30% (Razzell 1980). There is a possibility that in time it will be possible to adapt at modest cost these oil fired stations to burn a mixture of up to 40% fine coal and 60% oil; the technology, however, remains experimental and must be discounted for the time being.

It is possible to simulate, with reasonable confidence, the effects of different assumptions about future economic growth, alternative coal : oil price relationships and government policies upon the generating Boards' oil and coal consumption. The considerable and necessary data are of course confidential to the generating industry, the NCB and the Department of Energy, and have not been made public. Nevertheless, it is possible to show, from an accumulating body of secondary evidence, the most important and sensitive influences that are likely to bear upon the coal industry's market prospects in the electricity sector – and to demonstrate something of their magnitude. For example, one key assumption is the future relationship of coal and oil prices. In its 'reference scenario' for the 1978 Green Paper, the Department of Energy forecast coal demand in Britain in 1985 at 88 million tonnes. This estimate was based upon an undeclared assumption that the coal : oil price ratio in that year would be 0.75. If the ratio is narrowed to 0.9, however, then the estimated coal burn of the generating authorities falls to 70 million tonnes, a reduction of 20% (DEn 1980c). This relationship (if not its size) should not be surprising, since for some time the Department (1978a) have recognised that:

'The only significant component of the energy economy which appears to respond sensitively to fuel price movements is the electricity generating sector which derives its flexibility from the margin of capacity required to guarantee electricity supply during the daily peaks of demand' (p. 4).

What *was* surprising, perhaps even remiss, was the failure of the Department of Energy to conduct any sensitivity tests on the coal : oil price ratios used in their 1979 projection exercise – particularly when the relativities that were adopted were quite unique in the country's energy experience, widening from 0.6 in 1985 to 0.5 in 1990 and 0.4 in 2000.

However, the 1979 projections of the Department did have the compensating virtue of demonstrating the effects of variations in a second key assumption upon estimated electricity coal demand, the rate of future economic growth. Besides the published cases of 2.0% and 2.7% GDP growth per annum with estimates of a coal burn within the range 89 to 94 tonnes in 1990 (DEn 1979a), the Department also examined the implications of a 1.0% per annum growth in GDP upon electricity coal requirements – and found them to be as low as 80 million tonnes in the same year (DEn 1980d). Similarly, in evidence to the Vale of Belvoir Inquiry the CEGB (1980b) estimated that, in the year 1990–91, their coal burn would stand at 86 million tonnes assuming a coal : oil price ratio of 0.71 and a GNP growth rate of 2.5% per annum; however, under a growth rate assumption of 'between 1% and 1.5%', and the same coal : oil price ratio, coal demand was judged likely to be 74–9 million tonnes in the same year. This latter range of demand possibilities reflects yet a third key assumption in estimating future power station coal requirements – the level of installed nuclear capacity. The lower and upper figures assumed respectively 9500 MW and 7500 MW of installed nuclear plant.

In the absence of an ability, independently and with accuracy, to simulate the future coal demand of the electricity generating industry under different central assumptions – for want of the necessary facts and relationships – two courses are open to the independent observer. One is to attempt a crude simulation of the electricity industry's primary fuel demand from the limited number of facts and relationships that are known. The other is to derive from estimates and statements made by the electricity generating authorities and the Department of Energy some ordered measure of the effect of changing central assumptions upon forecasts of electricity coal demand.

At the Vale of Belvoir Inquiry, Robinson (1980) presented the findings of his independent, relatively simple but (when compared with official forecasts having access to more data) clearly robust model of primary fuel demand by the electricity generating industry. On its two most favourable assumptions from the viewpoint of coal sales (price ratios of 0.8 and 0.9), and assuming economic growth in the 1980s to be within the range of 2.0 to 2.5% increase in GDP each year, Robinson estimated *inter alia* a demand for 67–71 million tonnes of power station coal in 1985, and 68–73 million tonnes in 1990. Other evidence also suggests that the demand for power station coal in Britain could fall from 89 million tonnes in 1979 to less than 80 million tonnes in the mid-1980s, with little prospect for any significant growth in the rest of that decade (Table 4.2). There are, however, conflicting forecasts, and it is instructive to examine the differences in some detail before establishing a view on the most likely outcome. (In the data that follows, it has been necessary to convert some published data for England and Wales into a national equivalent; the assumptions that were made concerning coal burn in Scotland in different years are set out in Table 4.2).

First, and with reference to *the mid-1980s*, it is noteworthy that information supplied by the Department of Energy to the IEA in Paris, and published in 1978, estimated a fall in the electricity coal burn in Britain to 72.5 million tonnes by the year 1985. (The figure used by the Department in the 1978 Green Paper, by contrast, was 88 million tonnes.) Since the latter was based upon an assumed GNP growth rate of 3% per annum and

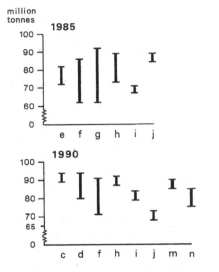

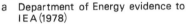

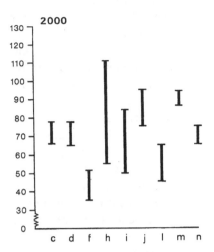

a   Department of Energy evidence to IEA (1978)

b   Department of Energy, Energy Policy (1978), reference case

c   Department of Energy, Energy Projections (1979)

d   Department of Energy, Energy Projections (1979) incl. 1% growth case

e   CEGB Corporate Plan (1978)

f   CEGB (August 1979)

g   CEGB evidence to House of Lords (1978)

h   CEGB (8 February 1980)

i   CEGB (28 February 1980)

j   National Coal Board et al. (Tripartite Group 1977)

k   National Coal Board (1979)

l   Robinson (1980)

m   WOCOL (1980)

n   This study (1980)

Figure 4.4   Alternative forecasts of power station coal demand in the UK in 1985, 1990 and 2000.

a coal : oil price ratio of 0.75, the lower figure submitted to the IEA must have assumed either very much slower economic growth (perhaps 1.5–2.0% per annum) or a much narrower price ratio (of perhaps 0.88) – or some combination of both. The assumptions have never in fact been made public.

The CEGB in their 1978 *Corporate plan* (using GNP growth rate assumptions of 2.0 and 3.2% in Cases I and II respectively) indicated that their coal burn in the mid-1980s could stand somewhere between 72 and 82 million tonnes, with a mid-point of 77 million tonnes. In evidence to the House of Lords' European Communities Committee in the same year, the Board forecast a coal burn in 1985–6 within the wider range of 62–92 million tonnes; once again, this had a mid-point of 77 million tonnes. Their August 1979 view, incorporated in answers to questions posed before the opening of the Vale of Belvoir Inquiry (CEGB 1979), was somewhat different; it estimated that coal demand in the mid-1980s would lie within a narrower and lower range – 62 to 86 million tonnes, with a mid-point of 74 million tonnes. During the course of the Inquiry itself, however, the Board first proposed on 8 February (CEGB 1980a) a higher central demand estimate for 1985–6 of 82 million tonnes (and a range of 73–89 million tonnes with slight variations in its future primary fuel price assumptions). However, three weeks later on 28 February (CEGB 1980b), following a further review of prospects, the Board lowered their central estimate again to 79 million tonnes. These changes in the CEGB's coal demand forecasts reflected shifting assumptions with regard to the coal : oil price ratio and the future rate of economic growth. The 28 February estimates also reflected revised expectations concerning both AGR capabilities and the amount of oil fired capacity that would be installed in 1985. All the forecasts however, embodied the expectation of lower power station coal consumption in the mid-1980s than the 89 million tonnes used in 1979 (Fig. 4.4).

Turning to the situation at *the end of the 1980s*, the CEGB in (August) 1979, estimated that their coal burn in 1990–91 would lie within the range 71–91 million tonnes, with a mid-point of 81 million tonnes. The 8 February 1980 estimate then raised expectations substantially to 87–92 million tonnes, with a central estimate of 91 million tonnes. The revised 28 February 1980 figures, on the other hand, lowered the range to 79–84 million tonnes, with a mid-point of 82 million tonnes, a reduction of between 8 and 12 million tonnes. Once again, the changing forecasts reflect the use of different basic assumptions. However, it is noteworthy that in all three forecasts (as in the Department's 1979 range of 80–94 million tonnes, the 1980 WOCOL projections and the much lower estimates of Robinson), the expectation is that there will be a slight increase in power station coal burn in the late 1980s. There are two principal reasons why this should be so, even under the assumptions of the CEGB's (lower growth) Case II August 1979 estimates, in which coal burning capacity is assumed to remain substantially unchanged. The first centres on the Board's (and the Department's) assumption that the coal : oil price ratio will widen between 1985 and 1990 – an assumption that is perfectly proper, yet one that is nevertheless at odds with the Electricity Council's interpretation of relative primary fuel price prospects that was noted earlier. The other is the expected decommissioning of some of the earliest Magnox nuclear stations and the completion of the new, large coal fired station at Drax in Yorkshire.

Table 4.5 Alternative forecasts of electricity coal burn in Britain in 1985, 1990 and 2000, and their key assumptions (million tonnes).

Coal : oil price ratio	Source	1985 GDP annual growth rate (%)				1990 GDP annual growth rate (%)					2000 GDP annual growth rate (%)				
		1.5±	2.0	2.5±	3.0±	1.0	1.5±	2.0	2.5±	2.7±	1.0	1.5±	2.0	2.5±	2.7±
1.1	3	—	60	62	—	—	—	61	65	—	—	35–55	54	61	67
1.1	2b	62	63	66	—	68	—	65	68	—	—	—	62	65	70
1.0	3	—	62	66	—	—	—	65	68	—	—	—	56	63	70
0.94	2a	—	67	69	86	—	79–84	68	72	—	—	—	59	66	74
0.9	3	—	—	—	—	—	—	—	—	—	—	—	—	—	—
0.9	1a	—	—	73	70	—	—	—	—	—	—	—	—	—	—
0.89	2	—	—	—	—	—	—	71	—	91	—	—	—	—	—
0.87	2b	—	68	71	—	—	—	70	73	—	—	61	68	76	—
0.8	3	79	—	—	—	—	—	—	—	—	—	—	—	—	—
0.8	2c	—	—	82	—	—	—	—	87	—	—	—	—	—	—
0.8	2b	—	—	—	88	—	—	—	—	—	—	—	—	—	—
0.79	2b	—	—	—	—	—	—	—	—	—	—	—	—	—	—
0.75	1a	—	—	—	—	—	—	—	—	—	—	—	—	—	—
E.t.m.o.b.*		80	85	88	90	80	82	89	92	94	59	79	70	82	98
0.72	2b	—	—	89	—	—	—	—	—	—	—	—	72	82	—
0.71	2a	—	—	—	—	—	—	—	91	—	52	—	—	—	35
0.71	2c	—	—	—	—	—	—	—	92	—	—	—	—	—	35
0.71	2b	—	—	—	—	—	—	—	—	—	—	73–84	—	—	—
0.65	2b	—	—	—	—	—	—	—	—	—	—	—	—	—	—
0.59	2c	—	—	—	—	—	—	—	—	—	—	—	—	66–77	87
0.59	2b	—	—	—	—	80	—	89	—	94	—	—	80	—	—
0.5	1b	—	—	—	—	—	—	—	—	—	—	62	—	100–110	—
0.4	3	—	—	—	—	—	—	—	—	—	—	—	70	86	—
0.4	1b	—	—	—	—	65	—	—	—	—	66	—	80	80	78
Assumed nuclear capacity (GW)		10–12				10–12					20	26+	31+	25–20	30–26

Source: 1, Department of Energy (1a, VBPI evidence regarding Green Paper: 1b, VBPI evidence regarding 1979 Projections).
2, CEGB (2a, 1978 *Corporate plan*; 2b, 1979 VBPI evidence; 2c, 1980, letter dated 8 February 1980; 2c, 1980, letter dated 28 February 1980).
3, Robinson (1980), year 2000: scenarios 1 and 3.
Other assumptions: SSEB coal burn 1985 – 7 million tonnes; 1990 – 5 million tonnes; 2000 – 5 million tonnes. Scotland: installed nuclear capacity in year 1990: 2.5 – 3.0 GW: in year 2000: 5 – 6 GW.
*Estimated technical minimum oil burn (0.74).

Agreement about the trends of power station coal demand in the earlier and later 1980s still leaves unresolved the question of its most likely magnitude. As has been seen, the range of the forecasts or estimates of demand for any one year is considerable. To bring some order to the plethora of figures already noted and to others found elsewhere, Table 4.5 demonstrates something of the relationship between the three key assumptions behind alternative electricity coal demand forecasts – the future coal : oil price ratio, the assumed rate of economic growth and the contribution of nuclear power. For the years 1985 and 1990 only, a narrow range of nuclear generating capacity is assumed; a much wider set of possibilities underlie the estimates of the subsequent decade. There are, of course, other (usually undeclared) assumptions lying behind all these figures – such as the relationship between the growth of GDP and the growth of electricity demand. These explain some of the smaller inconsistencies within the matrix, but do not justify further elucidation at this juncture. It should be noted in parenthesis that the 28 February 1980 estimates of the CEGB are not strictly comparable with the others to the extent that the forecasts were made after it came to be recognised that there would be somewhat smaller oil and nuclear generating capacities available than had previously been assumed. The coal burn in these estimates was consequently declared to be some 5 million tonnes higher than it would otherwise have been. It should also be noted in connection with Table 4.5 that there is a 'technical minimum' level of fuel consumption by those oil fired stations that were designed for base and middle load operation – it was pointed out earlier that these very large oil fired generating plants cannot be switched on and off at will – and this level is reached when the coal : oil price ratio is 0.74 (Razzell 1980). In consequence, any further widening of the ratio beyond 0.74 would not generate any significant increase in coal demand, *ceteris paribus*. Hence the 'break' in the table at this level.

Although it would be highly desirable for independent observers to have a full rather than a partial set of coal demand estimates in the matrix of Table 4.5, additional evidence is not publicly available; even the Inspector at the Vale of Belvoir Inquiry had access to no additional forecasts of electricity coal demand. The figures that are available, however, do fall into a meaningful pattern. For example, they reveal fairly convincingly that, in order to estimate with good reason a demand for power station coal in excess of (the 1979 level of) 89 million tonnes in either 1985 or 1990, it is necessary to assume (a) a rate of national economic growth of some 2.5% per annum throughout the whole decade, and (b) a coal : oil price ratio at or wider than 0.75 in 1985 and 1990. In the light of British economic performance in recent years and decades, and of coal : oil price relativities since 1973, these appear to be very unlikely joint prospects and hence (from the viewpoint of the NCB) somewhat optimistic assumptions.

It seems highly improbable that, for the forseeable future, economic growth in Britain will greatly exceed the annual rate of 2.1% that was achieved from the mid-1960s to the mid-1970s (Table 2.1). Even the 1978 assumptions of the IEA and the OECD – at 2.1% in the late 1980s, for example – must now be regarded as optimistic. For the purpose of this study however, the view is taken that the most realistic initial basis for market analysis is the assumption of a national growth of GDP averaging between 1.75 and 2.25% per annum, and with a central estimate of 2% per annum, over the next two decades.

Turning to the coal : oil price ratio, it seems highly unlikely that it will widen to anything like the differentials proposed by the Department of Energy for the period to the year 2000 (Table 4.3); and reservations must be expressed about the Generating Board's assumptions for the 1990s. Oil prices are likely to rise in real terms, certainly; but different views can be taken about the rate of increase. At the same time it cannot be escaped that the British coal industry has been, is and will remain, subject to very powerful upward pressures on its costs. Their prospective magnitude defies a scientific calculus, but their three principal components are clear. The first is miners' wages, these have increased substantially in real terms since 1972, and very much faster than average gross weekly earnings in the country as a whole. Between April 1973 and April 1978, earnings in all industries and services rose by 112%; in coal mining they rose by 158% (data from *New earnings survey*, 1973 and 1978). There is no reason to believe that this relative improvement in miners' wages will not be maintained, and it could well be increased in the next decade or more. If there is a common denominator in the international experience of the mining industry with respect to relative wage rates, it is that miners' wages fell relative to general industrial wages during the period of falling mining employment in the 1960s; but in the 1970s somewhat contrasting experiences were recorded. In the United States, an increase in the demand for mining labour was paralleled by rising relative wages. In West Germany, a continuing decline in mine employment was accompanied by substantially unchanged relative wages. Yet in Britain, a slight fall in colliery employment was accompanied by a significant relative improvement in miners' pay, a feature that can only be explained by the distinctive monopoly position of the NUM – and the political position of the miners after the 1972 and 1974 strikes. There is no evidence to suggest that – rightly or wrongly – the economic and political strength of the NUM is failing. If, therefore, the mining workforce continue to be paid well above the general industrial wage, and if the share of labour costs in the total costs of mining is relatively high (as it is in the case of underground operations especially), then the upward pressure on the costs of the industry will continue to be strong.

A second source of upward pressure on the coal industry's costs is the expense of mining equipment and supplies. This also appears to be rising both rapidly and in advance of inflation. The price of replacement mining machinery, roof supports, cutters, loaders, conveyors and other handling equipment, of maintenance spares and ancillary parts, has risen dramatically in recent years. As a measure of this, it can be noted that the costs of the 1974 *Plan for coal* (which in substantial measure is a programme of modernistion and re-equipment rather than new mine construction), rose by 43% in real terms between 1974 and 1979. Yet a third upward cost pressure is the financial charges that have to be carried by the coal industry. These have been growing rapidly in recent years – between 1974–5 and 1978–9 the NCB's annual investment trebled in real terms – and will undoubtedly be considerably more burdensome in the 1980s as the full impact of the industry's capital investment programme is felt. Although the provision in the 1980 Coal Industry Act, which allows for the deferment of interest payments in the case of large capital projects with long construction times, will ameliorate the situation to some degree, the mounting financial charges of the Board cannot be escaped in any assessment of its future.

Together, therefore, these three component costs – wages, equipment and materials, and financial charges – will exert considerable upward pressures on the mining industry's overall costs in the next two decades. This will severely limit the advantage that coal costs and prices might have over oil prices; it appears likely that they will increase steadily in line with world inflation, and quite possibly rise gradually in real terms. Certainly, there is no convincing evidence that, in relation to the probable trends in oil prices, the coal industry's costs will not follow the same path that they did during the period 1974 to 1979, and thereby severely limit the pricing headroom available to the NCB. For these reasons, this study takes a view similar to that expressed in 1979 by the Electricity Council, as noted earlier, and it assumes that for the foreseeable future the coal : oil price relationship in the power station market is likely to fluctuate between 0.7 and

0.9. The first figure is a slightly wider ratio than the 0.71 pertaining in mid-October 1979 when the effects of the Iranian crisis were feeding through into the markets for oil. However, price relationships existing on any particular date are a less useful guide in these matters than annual averages. The most recent annual figure of the CEGB, for the financial year 1978–9 was 0.88. The widest ratio was 0.65 in 1974–5 immediately after the four-fold increase in crude oil prices (Table 4.3).

Within these declared assumptions, Table 4.5 indicates that forecasts of power station coal demand in 1985 and 1990 vary much more with changes in the assumed coal : oil price ratio than with differences in the assumed economic growth rate. A central estimate for 1990 (taking into account the CEGB's most recent expectations of plant availability) would be 75 million tonnes. However, the possibility of a wider price ratio than 0.8 in the forecast year prompts the adoption of the range 75–85 million tonnes, the wider ratio being associated with a slower economic growth rate, in the subsequent estimates of total coal demand in Britain (Table 7.1). This range is somewhat wider than, but not substantially different from, the 79–84 million tonnes derived from the Generating Board's latest (late February 1980) estimates.

Uncertainties increase substantially in any consideration of the subsequent decade – *the 1990s*. Once again this is amply demonstrated in previously published figures (Table 4.1 and Fig. 4.4). The Tripartite Group in 1977, for example, suggested that there would be a power station market for 75–95 million tonnes of coal in the year 2000. In the 1978 Green Paper, the Department of Energy suggested a much smaller market of 64 million tonnes. The 1979 evidence of the NCB to the Commission on Energy and the Environment, on the other hand, adopted a figure of 90 million tonnes. But the Department (in its 1979 projections) again took a less optimistic view by proposing a market for 66–78 million tonnes. In the August 1979 opinion of the major purchaser of these coals (the CEGB), all these figures are too high; a demand range of 35–52 million tonnes (including 5 million tonnes for Scotland) was proposed. The 8 February 1980 estimates of the Board suggested a somewhat higher and much wider range of possibilities, stretching from 55 to 111 million tonnes; at its upper and lower ends this estimate demonstrated the effects on coal demand that would follow from highly contrasting scales (15 GW and 40 GW) of installed nuclear capacity at the turn of the century. The CEGB estimates dated 28 February 1980 reflected a more pessimistic interpretation of the country's economic prospects: they lowered and narrowed the forecast range once again, this time to 50–84 million tonnes, with a mid-point of 67 million tonnes. Compared with those of 8 February, and assuming 20 and 30 GW installed nuclear capacity, this new estimate indicated a reduction of 27 million tonnes in expected power station coal requirements in the year 2000 – a tonnage equivalent to the proposed output of nearly *four* Belvoir coalfields.

All of these forecasts, of necessity, employ certain assumptions concerning the relationship between the growth of the country's GDP and the increase of electricity demands. Yet the further a forecast reaches into the future, the more uncertain that relationship becomes. Many factors besides economic growth influence the consumption of electricity – social trends, the price of electricity and competing fuels, and developments in energy-using technologies, for example. Some of these factors are outside of the control of the electricity industry; others, such as tariff policies and marketing strategies, are obviously within the industry's grasp. In total, however, the future of these many influences is highly uncertain – an uncertainty which tends to be left unquantified in government and industry estimates of likely sales.

Whatever the level of electricity demand in the 1990s, it is clear from industry statements that both the CEGB and the SSEB would prefer to rely for its satisfaction upon the installation of a substantial number of nuclear generating plants. Their preference is based upon a belief, which is shared by the Department of Energy, that in the medium and longer term the costs of nuclear power are likely to be lower than coal fired generation costs, and moreover are *less uncertain* than the costs of coal fired generation. This analysis is based in turn upon a recognition that the key variables in nuclear costs are those of construction and plant performance, and not fuel cycle costs; in contrast, the most important element in coal fired stations, the obvious alternative, is the cost of coal and not construction costs or plant performance (Department of Energy 1978). The generating authorities have taken the view, therefore, that the costs of station construction and operation are less uncertain (and more subject to their own control) than are the future costs of purchasing primary fuel. Their nuclear ambitions were most fully reflected in the CEGB's 1978 *Corporate plan*. Under its assumptions, nuclear generating capacity in England and Wales alone would need to be increased from 15.6 GW in 1990 to a highly optimistic 79.5 GW in 2000 under the exceptionally high economic growth rate (3.2% annually) assumed in their Case I; and from 6.8 GW to 32.8 GW in their Case II. Meanwhile, the Generating Board suggested that the capacity of its coal fired plant in the 1990s would decline from 44.2 to 20.4 GW in Case I, and from 37.8 GW to 26.7 GW in

Case II. The implications of these trends for the CEGB's consumption of coal are dramatic. On the basis of these two somewhat extreme cases (and neglecting Scotland for the moment), the demand for power station coal would fall from 66–86 million tonnes in 1990, to 57–60 million tonnes in 1995, and then to 30–47 million tonnes at the turn of the century. The slower rate of decline in coal demand, incidentally, is associated with the lower level of economic growth and the less optimistic assessment of nuclear costs in Case II. Even more noteworthy is another assumption that informs these figures: this is that the coal : oil price ratio widens during the 1990s from 0.78 to 0.71 in both cases.

Once again, as for the previous decade, the numerical relationship between assumptions about economic growth, coal : oil price relativities and the demand for power station coal can be crudely demonstrated. Together with associated assumptions about nuclear capacity at the end of the century, these are set out in Table 4.5. From it three principal conclusions can be drawn. First, on any single set of assumptions covering the whole period 1980 to 2000, there is either no growth or a gradual decline in the demand for power station coal in the second decade – provided economic growth is at or below 2.25% per annum, and at least 25 GW of nuclear generating capacity are installed by the end of the century. Second, to estimate year 2000 requirements much in excess of, say, 85 million tonnes would require very bold assumptions indeed – growth rates sustained at or above the level of 2.5% per annum for two decades; coal : oil price relativities at or wider than 0.74 (the technical minimum oil burn) at the end of the period; a badly delayed (or cancelled) nuclear programme; or a combination of all three. The likelihood of such a set of circumstances is relatively remote. Third, given the probability that energy and electricity coefficients in the late 1980s and the 1990s will fall to lower levels than those assumed recently by both the Department of Energy and the Generating Boards, a central estimate of power station coal demand in 2000 would appear to be 70 million tonnes. This is based upon the assumptions (as before) that economic growth in the 1990s will lie between 1.75% and 2.25% per annum, that coal : oil price relativities will range between 0.7 and 0.9, and that 20–25 GW of nuclear capacity will be 'on stream' in that year.

The Government's plans for nuclear power station construction currently envisage at least 22 GW by the late 1990s (Table 3.1). The possibility that on the one hand this programme might be accelerated, but that on the other hand there might be a wider coal : oil price ratio in 2000 than 0.8 (plus an associated growth rate of less than 2% per annum), however, prompts the adoption of 65—75 million tonnes as this study's estimated range of power station coal demand at the end of the century (Table 7.2).

Looking beyond the year 2000, the CEGB in their 1978 *Corporate plan* took the view that their 'plant mix will include a proportion of coal fired plant which could result in coal consumption of more than 30 million tonnes per annum depending upon the relative economics of nuclear and fossil fuel plant when operating at less than base load' (p. 37). The Department of Energy have also taken the view that a decline of coal markets for power stations in the 1990s will continue into the 21st century. The magnitude of the decline may be in dispute; but an expected tendency for the market to continue contracting after the 1990s is undoubtedly significant for the longer-term planning of the British coal industry.

5 Markets for other than power station coal in the 1980s and 1990s

Outside the power station market, the most distinctive set of demands for coal stems from the iron and steel and foundry industries. Their requirements for coking coal contracted from 29 million tonnes in 1960 to less than 15 million tonnes in 1978 and 1979. It seems unlikely that this trend will be reversed. In addition to the problems posed by steel making over-capacity in the world at large, and the relatively high costs of many Western European producers, two factors will substantially influence the future size of the market for coking coal in Britain. One is the speed with which the British Steel Corporation (BSC) can rationalise its production of pig iron into fewer but larger units in order to reduce its unit costs. The second is the rate at which modern blast furnace practice, with very much lower coke rates (i.e. coke consumption per tonne of pig iron produced) than are currently recorded in Britain, can be adopted by the Corporation. Both factors will help to ensure the survival of a significant and viable British steel industry; but both factors, even with less oil and gas underfiring following price increases, will also reduce the consumption of coke in relation to steel demands. For the foreseeable future, and certainly throughout the 1980s, even with some continuing sales of foundry coke, the demand for coking coal would appear likely to decline, more particularly now that the BSC has announced a decision to reduce its steel making capacity from 21.5 million ingot tonnes to no more than 15 million ingot tonnes per annum as a result of deteriorating market prospects. The 1978 Green Paper forecast a demand for 24 million tonnes of coking coal in 1990. The Department of Energy's 1979 published projections reduced this figure to the range 15–18 million tonnes. Subsequently, the Department of Industry revised its official expectations of the future scale of the British iron and steel industry; with this in mind, and taking note of the Department of Energy's 1% economic growth case, the forecast for coking coal demand in 1990 fell to 10–12 million tonnes. This study adopts these figures, regarding the 1980 WOCOL estimate as either dated, or optimistic, or both (Table 5.1).

Table 5.1 Alternative forecasts of coking coal demand in Britain in 1990 and 2000 (million tonnes).

		1990	*2000*
1	Tripartite Group (1977)	—	20–25
2	NCB (1979)	—	20
3	Department of Energy (1978)	24	27
4	Department of Energy (1979a)	15–18	12–19
5	Department of Energy (1979b), including 1% growth case and revised	10–12	7–13
6	Robinson (1980)	—	13–15
7	WOCOL (1980)	17–20	15–20
8	this study	10–12	8–12

The 'other markets' for coal comprise sales to general industry, domestic consumers and a variety of smaller users. In total these 'other markets' purchased 25 million tonnes in 1979 – compared with 116 million tonnes in 1960. Outside the NCB there is general scepticism about any significant revival of these markets before the late 1980s. In the reference scenario of the Green Paper, a 1990 demand for 20 million tonnes was forecast; and in the Department's published 1979 projections, a range of 18–20 million tonnes was proposed and was further widened to 15–20 million tonnes under the 1% growth case (Table 5.2). In evidence to the Vale of Belvoir Inquiry, however, the Coal Board argued that general industry alone will provide a market for 15–16 million tonnes of coal by the mid-1980s, and that further growth can be expected in the latter half of the decade. The 1980 WOCOL study employed an estimate of 25–33 million tonnes of coal sales to 'Other markets' in 1990. These aspirations have to be weighed initially against the nature of the supporting evidence, and the views of the Department of Energy.

Coal sales in 1978 to general industry in Britain were 8.6 million tonnes. Sales in 1979 were 9.2 million tonnes. The NCB presented to the Vale of Belvoir Inquiry a schedule of prospective new business in the UK totalling over 5 million tonnes, which the Board expect will have been won by the middle 1980s. It is not at all clear how these gains will compare with

Table 5.2 Alternative forecasts of 'other markets' for coal in Britain in 1990 and 2000 (million tonnes).

Study	*1990*	*2000*
1 Tripartite Group (1977)	—	40–80
2 NCB (1979)	—	60
3 Department of Energy (1978)	19	74
4 Department of Energy (1979)	18–20	46–68
5 Department of Energy (1979) including 1% growth case	15–20	38–68
6 Robinson (1980)	—	22–35
7 WOCOL (1980)	25–33	48–80
8 This study	18–22	39–43

		Study (see table above)							
	(1979)	*1*	*2*	*3*	*4*	*5*	*6*	*7*	*8*
1990									
general industry (incl. collieries)	(10)	—	—	13	12–14	9–14	—	14–22	12–16
domestic and manufactured fuel	(13)	—	—	6	5	5	—	11	5
other consumers	(2)	—	—	1	1	1	—		1
SNG	—	—	—	—	—	—	—	—	—
total	(25)	—	—	20	18–20	15–20	—	25–33	18–22
2000									
general industry (incl. collieries)	(10)	30–50	40	41	39–45	32–45	15–25	28–47	32–36
domestic and manufactured fuel	(13)	10–30	20	3	3	3	7–10	16–20	3
other consumers	(2)			1	3– 5	3– 5			4
SNG	—			29	1–15	0–15		4–13	—
total	(25)	40–80	60	74	46–68	38–68	22–35	48–80	39–43

any decline of sales to existing customers which will follow from both the economic recession expected in the early 1980s and the introduction of conservation measures. The net gains might be only 4 or 3 million tonnes, which would leave the total industrial coal market at 12 or 13 million tonnes in 1985. Nor is it clear to what extent the 1979 and 1980 improvement in industrial sales is a function of the temporarily tight supplies of natural gas, and whether they will be sustained once additional North Sea supplies of gas become available. The NCB have not published an industrial sales forecast for 1990; the extent to which their views influenced WOCOL projections is not clear, although the Board were participating members of the study.

The Department of Energy's somewhat divergent views on the prospects for coal in the general industrial market in the 1980s are summarised in the 1978 Green Paper. There, it was stated that the premium energy requirements of industry in all likelihood will continue to be satisfied by natural gas, coke oven gas, burning oil and electricity; the non-premium market, on the other hand, will have a choice between interruptible natural gas, heavy fuel oil and coal. Interruptible gas, sometimes known as 'valley gas', is sold in order to limit the seasonal variations in the demand for gas and so to maintain a high load factor in the distribution system; priced at a discount, supplies can be withdrawn from consumers at relatively short notice. The Department took the view that the continued availability of interruptible natural gas could press industrial demand for coal down 'towards its lower non-substitutable limit of some 6 million tonnes (mainly for cement and aluminium)' in the 1980s before it started to recover once again. However, the Green Paper noted that 'Whenever natural gas supplies pass their peak as they eventually must, coal . . . could then begin to recover some of its (market) share'. Although the Department in 1978 saw interruptible gas supplies falling in the late 1980s, the gas industry itself is more optimistic about future supplies and, as has been noted, believes that it can maintain production at or above the 6000 m.c.f.d. level throughout the rest of this century – in effect an increasing quantity of useful energy as conservation measures take effect. In 1978 the Department of Energy forecast industrial coal sales of some 13 million tonnes in 1990. A range of 9–14 million tonnes emerges from their extended 1979 projection (Table 5.2).

This study adopts the range 12–16 million tonnes for industrial coal sales in 1990. The lower estimate reflects the 2% growth case of the Department, whilst the 16 million tonnes seeks to mirror something of the greater optimism of the Coal Board. Together, the 'Other markets' for coal are therefore estimated at 18–22 million tonnes in that same year, including 6 million tonnes of domestic, manufactured fuel and sundry additional coal sales (Table 5.2).

Much greater uncertainties lie ahead in the 1990s. The Department of Energy expects that domestic coal sales will continue to fall towards 3 million tonnes by the year 2000. The NCB, on the other hand, believes that its market in this sector will stabilize at 8 million tonnes. There is no ready means of adjudicating between these two propositions. It should be noted, however, that the Department's figure is set alongside their expectation of declining natural gas supplies in the 1990s; the NCB on the other hand have not declared their assumptions in this matter. There would appear to be growing agreement however, that in Britain (if not elsewhere in Western Europe) synthetic natural gas – SNG – production will be essentially 'experimental' before the turn of the century. In their 1978 reference scenario, the Department forecast that SNG demand for coal would grow from nil in 1995 to 29 million tonnes in 2000 (Table 2.4). However, in their 1979 projections, SNG demands are reduced to a maximum of 15 million tonnes, and only appear in the very high 2.7% growth case. It is noteworthy that the Secretary of State for Energy (1980a), in the second reading of the Coal Industry Bill, noted, 'Looking further ahead, possibly 20 years to the next century, rather than this . . . we can look forward to exploiting the potential of coal as a source of liquid and gaseous fuels.' It is certainly clear now that the British Gas Corporation see little need of commercial SNG production before the next century, a position that they impressed upon the Inspector at the 1979 Public Local Inquiry into the Beckton District Plan, where it was reported that the British Gas Corporation were 'still developing their coal gasification system at Westfield but assuming that the first full scale plant was required about 2000–2005'. The present study adopts the view of the British Gas Corporation, which is based in turn upon a more optimistic attitude towards prospective natural gas availabilities than that held by the Department of Energy. The expectation that natural gas supplies can be maintained at 6000 m.c.f.d. throughout the 1990s, and possibly increased above that level, also has implications for coal sales to general industry in that decade. This is the sector of the market which is subject to the greatest uncertainties.

In 1977, the NCB expressed its hope that there would be a considerable growth in the demand for coal by industry in general before the turn of the century. This hope was based upon their expectation of a 'substantial and sustained competitive margin' for coal; the promised advantages of small-scale fluidised bed combustion in the general industrial market; and the withdrawal of natural gas from non-premium uses (Tripartite Group 1977). Fluidised bed combustion is a means whereby coal is added to and burnt within a bed of particles, such as coal ash, sand or other refractory materials, which is maintained in a highly turbulent state by a flow of air; it considerably enhances the efficiency with which coal can be used, and permits the pollution-free burning of low grade coals, but it is still in the early stages of commercial application. The Coal Board's aspirations also rested substantially upon the assumption that the British Gas Corporation will by then have moved into a phase of falling production-to-reserves ratios, and an altered marketing strategy. The market forecast of the Board at this time was 30–50 million tonnes in the year 2000, a range that was narrowed to 40 million tonnes in their evidence submitted to the Commission on Energy and the Environment in 1979. The Department of Energy, in their 1978 reference scenario, estimated an industrial demand for 41 million tonnes, and in their 1979 projections they suggested a range of 32–45 million tonnes. The WOCOL projection of 1980 suggested an even wider range of 28–47 million tonnes. Robinson (1980), on the other hand, estimated a market as small as 15–25 million tonnes.

There are at least two reasons for such contrasting expectations (besides the adoption of markedly higher GNP growth assumptions underlying the upper forecasts of the Department and WOCOL). First, there is the disagreement about the future level of natural gas supplies; even to maintain production at 6000 m.c.f.d., let alone increase it above that level, would reduce the demand for general industrial coal on the Department's 2% growth assumption, from 39 million tonnes to perhaps 30 million tonnes. Robinson assumed gas consumption of 6000 m.c.f.d. at the upper end of his year 2000 range. Second, there are differences in forecasting methodology. The Department of Energy in their publication on *Energy forecasting methodology* (1978), have noted that forecasting the demand for industrial coal is particularly hazardous since:

'The energy sector . . . is characterised by many rigidities, lags and other imperfections and attempts to establish meaningful fuel price elasticities have so far proved unrewarding. The only significant component of the energy economy which appears to respond sensitively to fuel price movements is the electricity generating sector . . . ' (p. 4).

Robinson, on the other hand, made an estimate of future industrial coal sales with declared assumptions about future coal : oil price relativities, the growth of industrial production, and the rate of adoption of fluidised bed technology. Unlike those of the Department of Energy, which appears still to be using pre-1973 relationships in its forecasting model, his figures have the additional merit of being based upon more recent data and evidence. It could also be the case that the Department's 1979 assumptions about the country's future industrial structure give too great an importance to the manufacturing sector of the economy; and implicitly, the Department, like the NCB, could be placing too high an expectation upon the rate of adoption of fluidised bed technology. A realistic market analysis must also take fully into account coal's difficult handling characteristics, the problems of coal storage, consumer resistance to coal because of its associated dust and dirt, and the relatively high costs of coal transport by road and rail. In the years since coal served a substantial industrial market, the British railway system has been severely reduced, and the location of industry has become very much more geographically dispersed.

This study's assessment of the British industrial market for coal in the 1990s therefore can accept no more than modest growth from a narrow base – on present available evidence. It certainly would be incautious to assume that coal sales to general industry would increase at a faster rate than that proposed by the Department – 2 million tonnes per annum. On this *very generous* basis, the 1990 demand for 12–16 million tonnes would have increased to 32–36 million tonnes by 2000; are these figures built into the author's forecasts. In sum, the estimate of this study for all the 'other markets' for coal in the year 2000 (assuming, with the Department, a small rise in commercial and other sales to 4 million tonnes) is 39–43 million tonnes (Table 5.2). This represents between a three and a fourfold increase in sales to general industry, and 14–18 million tonnes of additional sales in 'other markets' for coal, between 1979 and the turn of the century.

These expectations are summed (later in Chapter 7), with the earlier estimates of demand by the electricity generating industry and by the steel and foundry industries. Next, however, consideration is given to the international aspects of the British coal industry, its export prospects and the challenge of imports.

6   *International trade in coal*

The market for coal in Britain is not necessarily identical to the demand for British coal; some British coal is exported while part of the UK market for coal is satisfied by imports. In recent years Britain has been a net importer of coal; although trade was in approximate balance in 1978, in the following year net imports were over 2 million tonnes once again.

Exports of coal have barely exceeded 2 million tonnes per annum since 1973. In 1978 and 1979 they stood at 2.3 million tonnes; yet the market which they most readily serve, Western Europe, has been increasing its coal imports steadily in recent years. The countries of the EEC, for example, imported some 30 million tonnes of coal in 1973; by 1979 these had nearly doubled to about 58 million tonnes. The origins of these Western European imports included Australia, Canada, Poland, South Africa, the Soviet Union and the United States. The reason for the NCB's inability to increase its sales into an expanding Western European market has at times been the result of inadequate supplies. More generally, however, the problem has been price, which in turn reflects the substantial difference between the costs of winning coal in Britain, and the delivered costs and prices of alternative suppliers overseas. Whilst the spot price of coal delivered to Western European ports in 1978 was around £15 per tonne, average NCB costs (excluding subsidies) in 1977–8 were nearly £22 per tonne. A third factor has been the reluctance of European buyers to enter into supply contracts with the NCB against the background of the miners' strikes of 1972 and 1974, the temporary ban that was then imposed upon British exports, and the fear of possible future interruptions to supply. As Lucas (1977) has noted, the European Commission has looked into the possibility of selling coal to North German utilities, but discovered that not only was British coal uncompetitive (which the Commission could do something about) but 'fundamentally was not considered a secure source of supply'.

Prospectively, these obstacles could be reduced. If the NCB proceed with, and succeed in, their production ambitions (see Chs 1 and 8) and if the earlier market analyses are correct, the supply constraint upon exports could be removed. The gap between the production costs of the British industry and the price of coal imports into Western Europe is unlikely to narrow (see below, p. 80), but it could be partly bridged by the energy policies of the EEC. The Commission has consistently argued since 1973–4 that it would be strategically and politically advantageous to encourage a greater use of coal in the Community, and to minimise the dependence of the EEC upon imported energy. Since 1977, therefore, the Commission has been regularly urging upon the Council of Ministers the advantages of providing a subsidy on the transport costs of steam coal entering intra-EEC trade. To date, however, the Council has consistently failed to agree upon the steam coal transport subsidy, and indeed there is a widespread belief even in the Commission that its implementation is most unlikely (Williams 1980). Consequently, whilst the British coal industry has benefited considerably from capital loans made available through the European Coal and Steel Community, it cannot realistically look to the rest of the EEC for significant coal markets. Even if the Commission's scheme were to be implemented, it is clear that the subvention proposed (some £6 per tonne in 1978) would not completely bridge the gap between average British costs and, say, Rotterdam prices. Moreover, it should be said that the tonnages that might be moved to the EEC and other Western European countries would be unlikely to exceed a few million tonnes each year in the 1980s, even if a substantial share of Eire's increasing power station requirements were to be satisfied by Britain. The NCB have suggested in the past the possibility of an export tonnage of up to 5 million tonnes each year. On present evidence this is an optimistic figure.

Imports of coal into Britain reached the level of 5 million tonnes in 1975. Subsequently they fell to just under half that level, then rose to 4.4 million tonnes in 1979. Some of the imports are coking grades, bought on long-term contracts by the BSC from Poland, the USA and Australia. Other imports are steam coal, ordered by the CEGB in the immediate wake of the coal supply difficulties that came with the 1973–4 oil crisis and the miners' strike in 1974. The probability is that the volume of coal imports will, at least in the short term, increase yet further.

Early in 1979, the BSC announced its need to import, in addition to its earlier supply contracts, approximately 0.5 million tonnes of Australian coking coal each year for its new Teesside blast furnace complex; the attraction of Australian coking coal lies in both its quality and its price. By November 1979, the BSC declared that it was prepared to double its imports to between 5 and 6 million tonnes per year (from a 1979 level of 2.9 million tonnes) unless NCB prices were reduced; such a level of imports could represent over half of the Steel Corporation's needs. At 1979 prices, imports represented enormous short term savings for the loss-making Corporation. The list price of coking coal from the NCB was about £40 per tonne, whilst an average delivered price of imported coal was about £30 per tonne. In fact the BSC claimed that £135 million out of its losses of £309 million in 1978/79 were the result of high domestic coke prices. Whilst direct and indirect subsidies of British coking coal cannot be ruled out, there can be little doubt that imports will tend to rise in the 1980s and only the speed of that increase is in doubt.

Also in 1979, and against the background of their relatively low stocks, the CEGB expressed a fear that in the forthcoming (1979/80) winter there could be a coal supply shortfall from the domestic coal industry. The Board therefore sought and received government approval to increase its imported supplies. In 1979–80, steam coal imports of some 2.5 million tonnes were expected. For 1980–81, a figure of 4.0 million tonnes has been quoted, and Razzell (1980) saw imports quickly building up to 5 million tonnes. Existing import facilities available to the CEGB, notably those on Thameside, can handle only *circa* 6 million tonnes each year. The Board, therefore, is currently searching for additional deep-water import facilities which would allow them not only to purchase up to 10 million tonnes of additional steam coal imports, but also to take advantage for the first time of the lower freight charges associated with vessels of (plus or minus) 100 000 dwt (deadweight tonnes) moving coal directly from its sources. With such deep water facilities, the CEGB's bargaining position with the NCB would be substantially strengthened, and the market for power station coal in Britain made markedly more competitive. The possibility of total CEGB imports of 15 million tonnes each year has been canvassed in a recent press report (*The Times*, 27 February 1980). It is noteworthy in this connection that, in its 1979 *Report* on the Area Electricity Boards, the Price Commission argued for increased coal imports – to hold down the CEGB's generating costs. It is also relevant to note that electricity coal is dearer in Scotland than in England and Wales; when the present planning agreement between the NCB and the SSEB runs out in 1983, therefore, it could well be advantageous to import steam coal into Scotland also.

Total British coal imports in 1980, could well exceed 5 million tonnes once again; unchecked imports in 1981 might be 7–9 million tonnes; and by the mid-1980s a figure of 12 million tonnes and over is possible.

The longer term relationship between the British energy market and the international coal trade is less easy to judge. It is inescapable that the British coal industry, like that in most of Western Europe, is primarily a deep-mine industry. Its situation contrasts acutely with a world coal industry that uses surface mine technology – with low labour inputs – for about half of its output today, and is judged by some observers likely to be over 80% opencast in the future (Shand 1978). Irrespective of geological conditions, deep mined coal is relatively expensive coal. In the geological conditions of Western Europe, deep mined coal is particularly expensive coal. The economic attractions of coal imports into Britain are, therefore, substantial and likely to increase.

It is significant that in West Germany the coal producers and the public utilities in 1980 signed a 15-year agreement under which the previous 10-year arrangement, for an average 33 million tonnes of indigenous coal each year to be supplied to public power stations, has been replaced by one which increases the tonnage to a maximum of 50 million tonnes in 1995. However, while all coal imports in recent years were limited to about 5 million tonnes per annum, the new agreement will allow the utilities to import one tonne of coal for every two tonnes of domestic coal burned up to 1987. After that date, the ratio will be one to one, allowing for a considerable increase in imports.

Table 6.1   Comparative international costs of mining coal, *circa* 1977 ($ US per tonne).

	Underground mining	(Source)	Surface mining	(Source)
France	36	(a)	—	
West Germany	47	(a)	—	
UK	43–47	(a)	32	(a)
Australia	18–25	(a, b, c)	8–24	(b, c)
India	10–16	(b)	—	—
South Africa	13–21	(a, b, c)	8–10	(c)
USA (Western States)	—	—	6–12	(d, e)

Sources: (a) Muir (1965); (b) Industrial Research Institute (1976); (c) IEA (1978); (d) Bureau of Mines (1976); (e) Mann and Heller (1978).

On average, British coal is cheaper to produce than West German coal. But its current production costs – an average of £21.75 per tonne in 1977–8, and £24.87 per tonne in 1978–9 – contrast vividly with (*circa* 1977) opencast costs of between £3 and £6 per tonne in such countries as

Table 6.2 Indicative steam coal costs and prices, delivered to NW Europe, 1979 ($ US per tonne).

From	Price FOB mine	Rail mine to port	Price FOB port	Port loading	Ocean freight	Port unload	Delivered price	Average CIF price
United States								
East, underground	20–35	10–15	30–45	1–2	6–10	2	39–59	49
West, opencast	8–18	10–20	20–35	1–2	8–11	2	31–50	41
Canada, West, opencast	15–20	10–20	25–35	1	8–12	2	36–50	42
Australia								
underground	15–25	5–10	20–25	2	10–14	2	34–43	39
opencast	12–20	5–10	18–25	2	10–14	2	32–43	38
South Africa, underground	10–15	5–7	15–22	1	8–10	2	26–35	31
Poland, underground	—	—	23–31	1	5	2	31–39	35

Source: WOCOL (1980).

South Africa, Australia and the United States (Table 6.1). To these production costs have to be added transport charges of perhaps £6–£10 per tonne (at 1977 prices) in order to land the coal in Western Europe. It is all too clear, however, that imported coals can be sold profitably in Britain at prices substantially lower than the domestic industry's average production costs. The CEGB have frequently noted that Australian coal, double handled via Rotterdam, is still cheaper than the domestic product delivered to the power stations on lower Thameside; and Razzell (1980) has confirmed that coal imports in 1980 were competitive as far inland as Didcot. With the provision of modern deep-water import facilities, the costs of transporting coal to Britain would fall by perhaps £2 per tonne, and the competitive position of imports would be considerably strengthened. Table 6.2 reproduces data first produced by the US Department of Energy which indicates something of the magnitude, price range and delivered price structure of steam coals delivered to North West Europe in 1979. The values in the right hand column have to be compared with average NCB pithead costs in 1978–9 of $57 (£1 = $2.30).

Although the future of imported coal is the subject of legitimate debate, several factors suggest that international coal prices are likely to increase at a slower rate than the NCB's production costs. It is worth quoting two short passages from the recent publication of the IEA on *Steam coal: prospects to 2000*:

'The coal industry, due to its relatively untapped resource base, is usually characterised as a constant cost industry, and coal supply curves are generally expected to remain relatively flat in real terms over the foreseeable range of coal demands likely to persist throughout the rest of this century . . . ' (p. 55).

In the same publication, the IEA go on to argue that the international coal industry is characterised by a geographical diversity of reserves, by vigorous competition in international markets, by a lack of corporate concentration, by the ability of new entrants to enter the market with ease, and hence by the probability that its medium term prices are unlikely to diverge significantly from its medium term costs:

'In the foreseeable future, therefore, coal prices are likely to be determined less by competition between coal and oil directly, and more by intra-fuel competition among coal suppliers and by competition between coal and nuclear power in the electricity generation sector . . . ' (p. 57).

The Department of Energy and the NCB, on the other hand, espouse the somewhat different logic that international coal prices will be pulled up towards international oil prices (which will continue to rise in real terms); that British oil will naturally be priced at international levels; but that British coal will enjoy the luxury of costs that increasingly diverge from oil prices (DEn 1980). The implication is that the coal industry will have increasing price flexibility, and will be able to take advantage of this in order to enlarge its domestic markets. Quite apart from any doubts there might be concerning the British coal industry's ability to contain its costs, this argument leaves unexplained why, with its prices implicitly lower than international coal prices, the NCB would not be tempted to export coal. It also leaves unexplained why international coal, which has a much more favourable cost structure than the British industry, would not be tempted to price itself down into the British market.

The cost and institutional structure of the international coal industry, and the prospective development of a substantial world trade in coal, in fact leads more readily to the conclusion that not only will other Western European countries continue to assert a preference for 'international' rather than British coal – and it is, of course, EEC policy not to restrict coal imports from countries outside the Community – but also that in the next decade British users other than the CEGB could also seek to take advantage of a low cost, non-oil, energy import. Their desire and interest in so doing seem likely to increase rather than decrease through time.

Small increases in British coal exports are just possible in the next two decades. Growing coal imports, for reasons of both quality and price, are rather more likely. In relation to home demands, however, and given the political position of both the NCB and the Mineworkers Union, it is highly unlikely that Britain's net imports of coal would be allowed to exceed, say, 10 or 15% of domestic coal requirements. On the basis of the earlier domestic demand estimates, this means that net imports are likely to remain below 10 million tonnes per annum in 1990, and 15 million tonnes in the year 2000. These are the figures that are adopted as upper limits to net imports in the summary of this study's estimates of the demand for British coal to the turn of the century that follows.

7 Market prospects: a summary

In the 1978 Green Paper on *Energy policy* it was proposed that Britain would afford a market for 129 million tonnes of coal in 1985 and 146 million tonnes in 1990. The more recent *Energy projections 1979* of the Department of Energy estimated somewhat poorer market prospects for coal in 1990, forecasting a demand between 124 and 132 million tonnes – with a mid-point of 128 million tonnes. Their 'modified' forecasts widen the range somewhat to 110–132 million tonnes, with a mid-point of 121 million tonnes; this is 25 million tonnes *less* than the reference case published in 1978. The Department's 1980 projections are awaited with considerable interest! The conclusion of the present study is that market prospects for coal in 1990 will be even poorer than those portrayed most recently by the Department of Energy, and that total demand in that year will stand no higher than 119 million tonnes and could be as low as 103 million tonnes; the mid-point of this range is 111 million tonnes (Table 7.1 and Fig. 7.1).

This finding is important in the subsequent evaluation of the NCB's production strategy through to the year 2000 (see Ch. 8). Moreover, by demonstrating that the markets for coal will be significantly smaller at the end of the 1980s than has hitherto been assumed by both the government and the NCB, certain implications follow for the analysis of coal market opportunities in the 1990s. It cannot be escaped that any growth in the demand for coal in that decade will begin from a much lower base than Coal Board planning has hitherto expected; and that lower base provides an important constraint upon the development of further market opportunities.

Table 7.2 summarises, and Figure 7.1 illustrates, the even starker contrasts in the forecasts of coal demand in the year 2000 by the Tripartite Group in 1977, by the Department of Energy in 1978 and 1979, by the NCB in early 1979, by Robinson and WOCOL in 1980, and in this study.

Table 7.1 Alternative forecasts of the total British market for coal in 1990 (million tonnes).

Actual			NCB	Department of Energy			WOCOL	This study
1978	1979			(1978)	(1979a)	(1979b)	(1980)	
81	89	power stations	?	103	89–94	80–94	85–90	75–85
15	15	coke ovens	?	24	15–18	10–12	17–20	10–12
24	25	other markets	?	19	18–20	15–20	25–33	18–22
120	129	Total	?	146	124–132	105–126	126–146	103–119
		(mid point)			(128)	(116)	(136)	(111)

Table 7.2 Alternative forecasts of the total British market for coal in 2000 (million tonnes).

Actual			Tripartite Group (1977)	Department of Energy (1978)	NCB (1979)	Department of Energy		Robinson (1980)	WOCOL (1980)	This study
1978	1979					(1979a)	(1979b)			
81	89	power stations	75–95	64	90	66–78	65–78	45–65	86–94	65–75
15	15	coke ovens	20–25	27	20	16–19	7–13	13–15	15–20	8–12
		other markets								
9	9	(a) industry	30–50	41	40	39–45	32–45	15–25	28–47	32–36
15	16	(b) domestic, commercial (incl. SNG)	10–30	33	20	7–23	6–23	7–10	20–33	7
120	129	Total	135–200	165	170	128–165	110–159	80–115	151–204	112–130
		(mid-point)	(168)			(147)	(135)	(98)	(178)	(121)

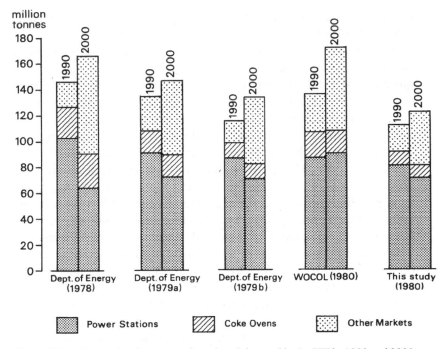

Figure 7.1 Alternative forecasts of total coal demand in the UK in 1990 and 2000.

On the best evidence available, and on the assumptions noted earlier, *inclu-ding the totality of recent and current (late 1980) government policies*, the most likely size of the market for coal in Britain in the year 2000 would appear to be between 112 and 130 million tonnes, with a mid-point of 121 million tonnes. It is necessary to italicize the words 'the totality of recent and current government policies', for it is just as important to recognise that government openly accepts the notion of 'consumer sover-eignty', and gives a considerable degree of discretion to the consumer (including the electricity generating boards) in the choice of different fuels, as it is to recall government's belief in 'a large and continuing need for coal.' Alter the assumptions about economic growth prospects, about future energy coefficients, about the coal : oil price ratio or about natural gas supplies, and the forecast demand range could possibly be higher. *Ceteris paribus*, however, only a significantly delayed – or a deliberately reduced – nuclear power programme than that which is the current ambi-tion of the electricity generating authorities and the Government could almost certainly ensure that the middle estimate of the year 2000 forecast range would be exceeded. Were that to happen, then demand for coal would be pushed 'at a stroke' up towards 130 million tonnes. Such an eventuality cannot, of course, be completely discounted.

Whilst markets for coal in the 1980s are substantially determined by investments already made or planned, the energy market in the subsequent decade is still very much subject to the effects of future government policy initiatives. Markets for coal in excess of, say, 120 million tonnes per annum undoubtedly could be secured in Britain through public decisions not only to forego some of the proposed nuclear power stations, but also to increase the coal burn of the electricity generating authorities through subsidies or edict. Capital grants for industries switching to coal fired boilers could assist demand in the same direction. Alternatively, government policies to slow down the depletion rate, and to increase substantially the market price, of North Sea gas might more indirectly serve the same end. The possibility of these and other policy decisions must be weighed in any final analysis of coal demand prospects in the last decade of this century. It is noteworthy that the Department of Energy in its *Energy projections 1979* recognised the need for 'some flexibility in *overcoming uncertainty in the timing of the development of new markets for coal*'. The Department suggested in consequence the replanting of older coal fired power stations; to date, however, no such decisions have in fact been taken.

It cannot be too strongly stressed, however, that there is little logic in proceeding with a major programme to expand this country's capacity to produce coal without simultaneously being assured with some confidence that there are uses to which that coal can be put. Particularly is this the case when the major investments required to ensure the ability to consume coal (new coal fired power stations, or major SNG plants, for example) take as long to construct as major mine modernisations and extensions. Until government proposes measures that are likely to ensure the existence of coal markets above the levels of the present study's forecasts, and these have been given expression in committed investments and firm plans, the case for adjusting coal production plans downwards towards a more realistic set of prospective market opportunities is very powerful indeed.

The possibility and the timing of an upturn in the demand for British coal are highly debatable. These questions are compounded by the relatively high costs of the British industry by international standards, and the probability of net coal imports tending to increase through time. Such additional uncertainties extend the forecast range of demand in 1990 and 2000 that is presented in this evidence. Table 7.3 summarises these findings,

Table 7.3 The demand for British coal in 1978 and 1979, and the author's estimates for 1990 and 2000 (million tonnes).

	Actual			
1978	*1979*		*1990*	*2000*
81	89	power stations	75−85	65−75
15	15	coke ovens	10−12	8−12
24	25	other markets	18−22	39−43
120	129	*Total home market*	103−119	112−130
		(mid-point)	(111)	(121)
—	2	*minus* net imports	0−10	0−15
120	127	*Total demand*	93−119	97−130
		(mid-point)	(106)	(114)

which indicate the prospect of a *falling demand* for British coal from 129 million tonnes in 1979 to a forecast range of 93−119 million tonnes in 1990, with a mid-point of 106 million tonnes in that year. Subsequently, demand could rise slightly. The forecast range for the year 2000 is 97−130 million tonnes, the upper figure assuming unrealistically that there will be no net imports in that year. Under present policies it is much more probable that the mid-point in the estimated demand range, 114 million tonnes, will be nearer to reality. It is against this interpretation of demand prospects that the production strategy of the NCB is reconsidered in the next section.

8 *The NCB plans reviewed*

The NCB's modernisation and expansion plans were initially forged against the background of very different energy and coal market expectations from those revealed in this study. A fundamental review is long overdue. Whilst it has to be accepted, along with Coal Board spokesmen, that the supply of coal is not something that can be turned on and off at will, and that there is a need to provide the industry with as stable an economic environment and production objectives as possible, such pleas cannot postpone indefinitely a realistic and critical examination of the production targets of the NCB.

The most recent statement of the Board's production capabilities and plans is that presented to the Commission on Energy and the Environment in 1979, and confirmed at the Vale of Belvoir Inquiry. It is summarised in Table 1.1, and raises as many questions as it answers. At the very least, a number of fundamental qualifications must surround their figure of future NCB production capabilities, particularly as they relate to existing mines. All of these qualifications suggest that the capacity figures quoted by the Board in its plans and public statements if anything *underestimate* the production potential of the industry in its existing and modernised facilities in the 1980s and 1990s. In consequence, and quite apart from the implications of the demand analyses presented earlier, they throw serious doubts upon the necessity for the industry to open as many new mines as the Board have proposed. They also raise questions about the size of the industry's capital requirements over the next twenty years.

First, it is not at all clear whether the Board's figures in Table 1.1 take into full and proper account the effects of any productivity increases that might be achieved at existing mines in the foreseeable future. The 1960s saw substantial improvements in British coal mining productivity following the introduction of powered supports and mechanised systems. The 1970s, however, have seen productivity increases slow, then disappear, and then move forward slowly again. Nevertheless, performance in

1979–80 (470 tonnes per man-year) was poorer than in 1972–3 (480 tonnes per man-year). These averages, however, have to be set against the fact that the top 5% of coal faces have recorded productivities of more than twice the national average, whilst some – including the new Advanced Technology Mining faces – are associated with output levels three times the national average (Davies 1978). There is little doubt that, technically, British coal mining productivity could be increased significantly through the introduction of further mechanisation and a fuller exploitation of existing capital investments in existing pits. In turn this higher output in relation to both capital and labour inputs would permit higher levels of coal production in existing pits than have been achieved there in recent years – even allowing for such technical constraints elsewhere in the mining system such as the capacity of winding gear or washeries. Were such productivity improvements actually to be achieved, then it seems likely that the capacity figures relating to existing and modernised mines in Table 1.1 would need to be adjusted upwards.

Second, it is arguable that quite modest changes in the standard mining practices of the NCB – such as reducing the thickness of the coal left in a seam after a cut, or leaving less coal in odd-shaped areas underground – could also increase substantially the output potential of existing facilities. Certainly, only about 50% of the coal *in situ* is extracted at mines currently in production, the main reason being the widespread use of mechanical longwall mining which is an inherently inflexible mode of coal recovery. It was noted earlier that the proposed Vale of Belvoir mines would extract only about 40% of the coal in place, and one plan for a new colliery at Park in Staffordshire would remove only 33%. However, the NCB have claimed that recovery rates will improve. They have even suggested that a return towards the 70–80% extraction rates that were often obtained at the turn of the century can be expected once again by the year 2000 (Tregelles 1976). At any given level of output in a particular mine, such an achievement would extend the length of time over which production was possible. More likely, however, improved recovery rates would allow some increase in the annual production of the Board's collieries and thereby once again require some increase in the capacity figures recorded in Table 1.1.

Third, it has been suggested by a senior NCB official that the output from the new pits at Selby could be higher than that originally planned, since the Barnsley seam at Wistow has been found to be thicker than was originally expected. The Board is naturally cautious about espousing a Selby production target in excess of the present 10 million tonnes per annum. Already that target represents a scale of operations that will be uniquely large in British coal mining experience – only a handful of mines currently produce more than 2 million tonnes each year – and there could well be major organisational difficulties underground, and possibly surface transport congestion, if additional output were to be attempted. On the other hand, the views of the mining engineers working on the sites cannot be completely ignored, and their public assertions that much larger tonnages could be won from the Selby field merit some attention (*Financial Times*, 4 May 1979).

Fourth, it was established at the Vale of Belvoir Inquiry that the NCB's future capacity at existing mines – noted as 105 and 80 million tonnes in 1990 and 2000 respectively in Table 1.1 – is somewhat greater than declared. This follows from the decision to reopen the pit at Thorne in Yorkshire, which will provide an extra 2.1 million tonnes of coal in both years, an output which is not recorded anywhere in Table 1.1.

Fifth, since coal reserves and the attractiveness of the exploitation is partly a function of energy prices, and since real energy prices are widely expected to increase in the future, it is more than likely that at least some new and profitable mining opportunities will be recognised in existing mines. The discovery of an extra 36 million annual tonnes of attractive mining opportunities in the existing pits of the NCB between *Plan for coal* (1974) and *Coal for the future* (1977) may have been exceptional. It nevertheless indicates the speed with which the production potential of existing mines can be revised in an era of rising energy values.

Sixth, and perhaps most important, there is the matter of pit closures. In its 1979 evidence to the Commission on Energy and the Environment, the NCB assumed that, through mine exhaustion, its capacity will fall by 3 million tonnes per year between 1985 and 1990, and by 2.5 million tonnes per year between 1990 and 2000. Previously, however, in *Coal for the future* the Board had taken the view that the loss of capacity on this score would average only 2 million tonnes per annum. Moreover, it is not at all clear that the Board's figures of future capacity have taken fully into account the difficulties that are likely to arise from the NUM who are reluctant to accept (and at times totally oppose) mine closures. It cannot be ignored that the NUM are in a position virtually to veto the closure of older and higher cost mines. In recent years it has forced the Board to retain for some time production capacity at Walton Colliery in North Yorkshire and (until the vagaries of geology prevailed) at Deep Duffryn in South Wales. More recently the President of the Union was reported as saying:

'I am not allowing any pits to be closed in Britain just because the Coal Board say they cannot sell the coking coal to the steel industry. I am not having what is happening in another industry determine the life of our pits, especially at a time when some customers are being allowed to import coal which is available in Britain' (*The Times*, 4 June 1980).

Setting aside the hyperbole of such curious economic logic, there can be little doubt that a substantial pit closure programme would involve either considerable redundancies or the inter-coalfield transfer of miners. Experience does not suggest that this will be easy, and indeed points to the fact that the NCB could well have difficulty in closing down between 2.5 and 3.0 million tonnes of older capacity each year between now and the year 2000. They may have to be content with the earlier figure of 2 million tonnes per annum.

In sum, the declared production capacity and output expectations of the NCB at their existing pits (including Selby), as summarised in Table 1.1, represent the least that might reasonably be expected in the light of the industry's committed and declared investment plans. Mining capacity from the middle 1980s until the end of the century could be several million tonnes per annum in excess of the figures quoted in the table. The extra output that might come from Selby (say, 2 million tonnes per annum), the known plans for Thorne (over 2 million tonnes per annum), the effects of higher productivity, altered mining practices, higher energy prices and slower pit closures could well represent up to 10 million tonnes of additional capacity overall.

There has to be set against these potential 'gains' in productive capacity, however, the harsh experience of mining realities. It has to be recognised, with Moses (1979), that production plans can disappoint and even fail, and that the past experience of the British coal mining industry is that output achievements tend to lie some 6–8% below the plans of the mining engineers. There also exist doubts concerning the ability of the industry fully to achieve its opencast production ambitions in the light of growing opposition from an increasingly articulate environmental lobby. In other words, the likely level of under-achievement in the modernisation and expansion plans of the Board could well offset the 'gains' noted earlier. These matters are imperfectly summarised in tabular form. Nevertheless, Table 8.1 sets out a somewhat fuller appraisal of the NCB's production capacity to the year 2000, without resort to new mining capacity in Leicestershire, Warwickshire or Staffordshire.

In the same table, these capacity figures are set alongside this study's

Table 8.1 Revised estimates of the production capacity of the NCB without further new mines, and the demand for British coal, 1990 and 2000 (million tonnes).

	1990	*2000*
Production capacity of the NCB		
declared NCB capacity without further new mines (Table 1.1)	129	110
plus extra output at Selby	—	2+
Thorne	2+	2+
Effects of higher productivity, changed mining practices, higher energy prices and slower pit closures	10±	9±
sub-total	141	123
minus production or capacity shortfall underground and/or lower opencast	10±	9±
Total capacity	131	114
Demand for British coal (Table 7.3)	*93–119*	*97–130*

earlier estimates of the demand for British coal in 1990 and 2000. The evidence gives cause for concern. It suggests that there could well be a substantial surplus of British coal mining capacity in 1990 – even if there are no net imports in that year. This surplus could lead to *either* the 'premature' closure of some mines and hence a reduced level of production capacity in 2000; *or* (if the mines were to be left open) rather greater reserves of coal in existing mines than is currently expected and hence a somewhat larger production capacity in 2000. These uncertainties set aside, the evidence of Table 8.1 also suggests that – assuming net coal imports of 10 million tonnes in 2000 – current, approved and identified capacity at existing mines (including Selby) could well cover market requirements at the turn of the century. The need for the NCB to develop additional mining capacity at new sites, such as in the Vale of Belvoir or at Park in Staffordshire, in consequence looks highly questionable.

The case for a review of the NCB's production plans, and especially its plans for new mines, does not stop at this point. The economics of exploiting new coalfields, even if the markets were readily available to utilize their output, are substantially less attractive to this country than the Board or successive governments have implied in public debate. It has been claimed by the Board, and indeed also in the 1978 Green Paper on *Energy policy*, that the total cost of coal won from new mines is likely to be lower than the operating costs of much existing capacity. The (1980) average cost of winning NCB coal was about £25 per tonne. This generalisation cloaks a wide range of mining circumstances and costs. Many of the Board's oldest mines, on the one hand, sunk near the turn of the century, meeting complex geological conditions, producing only a small tonnage each year and with very substantial unit labour requirements, must have costs in the region of £40 per tonne – or even more. The costs of winning Selby coal, on the other hand, could range between £15 and £18 per tonne (in 1980 values); its estimated capital costs of £71 per annual tonne (in 1978 values), the scale of operations, the drift technology, the ability to use the coal directly in local power stations, and various other factors all suggest that it ought to be a highly profitable production unit. But Selby could turn out to be a quite unique prospect, since many other new mines have direct and indirect costs that put them in a very different category. The three mines proposed by the NCB in the Vale of Belvoir, for example, look far from low cost. The direct costs of Belvoir coal could well be within the range of £19–22 per tonne, much depending upon the eventual expense of actually constructing the mines, and also upon the labour productivities actually achieved. Certainly it is likely to be considerably more expensive than Selby coal, given the shaft technology proposed for the Vale workings, the problems likely to be posed by underground water and the mines' relatively high capital costs – estimated at £105 per annual tonne (in 1978 values) by the NCB, and at £165 per annual tonne by the CEGB (Razzell 1980). This latter figure, it should be noted, is 2.3 times greater than the comparable figure for Selby. Indeed, the direct costs of Belvoir coal could well be higher than those of coal won from some of the modernised and reconstructed pits of the East Midlands and Yorkshire, where the capital costs of additional capacity range between only £1.50 and £10 per annual tonne (Moses 1979b). It is noteworthy that the Commission of the EEC (1980) recently commented, in connection with Community coal developments as a whole, that:

'Capital investment must be increased to raise capacities and achieve rationalisation. The advantage of modern and generously scaled layout for new mines is likely to be balanced out by the high capital outlay and the more difficult working conditions at greater depths' (Commission of the European Communities 1980 p. 15).

Direct cost comparisons between new and old mines, instructive though they may be, nevertheless cloak to some degree a wider economic and social reality. In addition to its direct costs, coal mined from the Vale of Belvoir, for example, could well have to bear the expense of long-distance waste disposal, the Board's proposals for tipping spoil adjacent to the mines being so widely unacceptable. Every 10 tonnes of material mined from the Vale mines is likely to comprise only 7.3 million tonnes of saleable coal plus 1.6 tonnes of run-of-mine dirt and 1.1 tonnes of development waste. The disposal of the waste in the clay pits of Bedfordshire, for example, could (in 1979 values) add £1.05–£2.68 per saleable tonne in transport charges alone to the direct costs of the coal. Moreover, the mining of coal in the Vale of Belvoir, as in all other new coal fields, would impose considerable costs upon the agricultural industry; it would demand considerable public investment in housing and infra-structure; it would occasion the loss of valuable recreational resources; and it would impose a wide range of environmental costs upon the surrounding communities. The complexity and something of the magnitude of these costs were exposed at the Vale of Belvoir Inquiry. Of course, these costs are not carried by the NCB, nor are they taken into account by the Department of Energy in the normal course of public investment appraisal unless they are an identifiable charge on the project. The costs nevertheless exist and have to be carried by someone.

At the very best, it can be fairly claimed that, if the NCB were required to pay the full 'external' or social costs of its proposed development in the Vale of Belvoir (on the well-established principle that the 'polluter pays'), the denial of mines there would certainly not deprive the country of an outstandingly cheap source of energy. Indeed, the evidence suggests that Belvoir coal could well be relatively expensive coal and, adding a notional figure for 'external' or social costs (plus some consideration of the British propensity in recent decades to lose control over the costs of large domestic construction projects), it would not be much cheaper than the direct cost of the average tonne of coal produced in Britain today. Certainly very much cheaper coal in terms of its marginal costs could well be won by the modernisation of many existing mines, for which the social costs have already been substantially paid. The route towards the cost-effective modernisation of British coal production is not as simple as the NCB and the Department of Energy have tended to imply in their public relations and official publications.

9 *The present coal question*

The coal industry's problems in Britain are daunting. By international standards, the country is a high cost producer. The NCB needs to move as quickly as investment, its workforce and social considerations will allow towards higher levels of productivity and lower unit costs. This can best be achieved initially through the completion of Selby and the extension of plant and facilities at its lower cost mines; and through the closure of pits in the higher cost coal fields. In general the NCB has had, and retains, the backing of public policies in these ambitions. The lack of growth in the markets for coal, however, severely limits the industry's ability to reach these objectives. Moreover, the Board has the full backing of the NUM for only one half of this policy, namely the investment in new facilities; and the Union has set its face against the early closure of some of the Board's high cost mines. Only by exaggerating its market prospects in the 1980s and 1990s has the Board been able to impose a logic upon its present investment programme, to justify its considerable claims on scarce (public) capital resources, and still to retain the cooperation of the Union.

No industry that invests on the basis of a false market prospectus can hope to yield a reasonable return to either its capital or its labour. An acceptable future for the British coal industry therefore must be sought within realistic market parameters. Given the progress made by the NCB since 1973 in the modernisation and expansion of existing mines and in the development of the Selby field, and given the investment programme already approved by the Board and the Government, now is the time when the NCB could well and with advantage pause and take stock. It should seek to consolidate its position upon its committed investments – whilst at the same time pressing upon its management and the NUM the necessity for productivity increases, cost reductions and additional mine closures in the 1980s, at least in line with the revised target of 2.5–3.0 million tonnes of capacity each year. If the closures can be accelerated to bring capacity more closely into line with demand and to allow some incentive for significant increases in capital and labour productivity at the remaining pits (increases that are unlikely to materialise in a period of surplus mining capacity such as the late 1980s could bring), so much the better. In such a context, the projects already approved, plus further investment at existing mines, seem likely to be more than adequate to satisfy market requirements to the turn of the century.

At some time in the future, no doubt, this capacity will begin to appear incapable of meeting prospective demands, and further decisions on additional mining capacity (including new mines) will be required. On the basis of the demand forecasts offered earlier, it could well be some time after the year 2000 that further capacity will be required – although the precise timing depends upon the country's rate of economic growth, the speed with which energy conservation measures are adopted, the achievements of the nuclear power programme, the productivity gains of the British mining industry, political attitudes towards imported coal, and the like. To provide new mining capacity after the turn of the century would demand planning initiatives in the middle or possibly early 1990s – but not before. By that time, the exploration and development programme of the NCB will doubtless have moved substantially forward. It will, therefore, have provided both the industry itself and the government of the day with some indication of the full range of mining options that are open to the Board in different places – plus some measure of the economic and social costs that might properly be attached to each of them. This is crucial information that was substantially denied to the Vale of Belvoir Inquiry, for lack of knowledge and the assertion of the NCB that an urgent decision to proceed with further new mines was required.

Within a few years, moreover, the costs of new mining ventures in Britain are likely to be somewhat clearer than they are today, and a proper evaluation of the Board's crucial experience at Selby especially, but also elsewhere, will be available to inform policy. The original estimate of investment under *Plan for coal* was £1400 million at March 1974 prices; a revised estimate at July 1979 prices is £5170 million; after allowing for inflation, this represented a real increase of 43%. A 1976 estimate of the cost of the Selby complex was £400 million; in 1978 the House of Lords was given a figure of £500 million; the Vale of Belvoir Inquiry was offered yet another estimate of £710 million (at 1978 prices). Not only could the real costs of new mines become much clearer in a few years time, but, given the revival of the fortunes of the coal industry world-wide, it is quite possible that improved or new technologies will have emerged to lower the costs of winning coal underground, to reduce the expense of waste disposal, and to improve the comparative cost position of the British coal industry.

It is always possible, of course, that the market forecasts for British coal which have been offered in this study will be erroneous to the extent that the demand for coal in the 1980s and 1990s will turn out to be somewhat higher than that which has been proposed, or that demand will begin to increase somewhat earlier than has been suggested. A substantial delay in the construction of the proposed additional nuclear power stations cannot be totally discounted, for example; nor can a sudden burst of high economic growth. In such eventualities, however, the major sources of coal demand are likely to be the electricity generating industry and, to a lesser extent, larger industrial users. Indications of the former's prospective needs tend to be known several years in advance of the coal actually being required, time enough for the NCB to make an investment response. Even in the general industrial market however, a substantial time-horizon exists within which to meet the needs of the larger potential users. The lead time for new mines is, of course, tending to get longer. An investment and production response to unexpected coal demands could be somewhat delayed therefore, if a reliance was to be placed solely upon new mines. However, any shortfall in coal supplies for two or three years could with no difficulty be made up with relatively low cost imports from a world market that is likely to be very much larger, and very much more sophisticated in its organisation, than it is today (IEA 1978). By 1990, it has been suggested, OECD Europe could be importing between 153 and 179 million tonnes of coal each year; and by the year 2000, Europe's imports could be in the 300–450 million tonne range. A few extra tonnes for Britain would impose very few strains upon such a supply system. It is surely better to pay that short-term import price for a few years than to carry a long-term burden of surplus and (internationally) high cost mining capacity. In any case, it makes little sense to commit major sums of capital to a nuclear power programme, and then simultaneously to invest in additional mining capacity in case the nuclear programme should fail.

Putting the matter another way, it is important in circumstances such as those facing the British coal industry to consider with care *the costs of being wrong* in the matter of market forecasts. If the estimates of the demand prospects for British coal that are proposed in this study prove to be too conservative and the production plans of the NCB had been adjusted downwards to meet them, then the country would face a short-term burden upon its balance of payments (but simultaneously could well enjoy the benefit of more lower cost coal imports than it would otherwise have used). If, on the other hand, the forecasts of the NCB and the Department of Energy are proved to be wrong, then the country will (literally) have sunk large sums of scarce public capital in, at best, a series of badly timed coal mining ventures, it will be faced with an embarrassing surplus of mining capacity, it will be required to close down rapidly and prematurely many relatively small pits in different parts of the country, and it will have to face the social disruption that such a course of action would entail. An over-investment by the BSC in modern steel plant in recent years has not assisted the cash flow and finances of that beleaguered enterprise, nor helped its prospects of successfully balancing the economic and social objectives it has always espoused. The approach to energy planning adopted by the Department of Energy, one which seeks to 'minimise regret' through overinvestment on the supply side of the national energy equation, has not been sufficiently appreciated and openly challenged in public debate. Some insurance premia are worth paying; others – including an unnecessarily large investment in coal – are not.

The time has come, therefore, to reconsider very carefully the national strategy towards coal and the British coal industry. The central question relates not to the industry's resource base, which is plentiful, if high cost; not to the industry's production capabilities, although these could well be more constrained than was once thought; but rather to the industry's market opportunities. In the light of a realistic set of assumptions about economic and energy demand growth, about competition in the markets for energy, and about public policies in the round, a fresh public appraisal of likely British coal sales over the next 20 years is required. Against this analysis the need is urgent *either* for new public policies to be forged that will ensure a higher volume of sales than currently seems likely, *or* for a major adjustment of the investment plans of the NCB to bring them more into line with future market probabilities. Manifestly, Britain does not require *both* the scale of investment proposed by the coal industry *and* a nuclear power programme of the size espoused by both the electricity industry and the government. And if the nuclear programme is to stand, the likely realities of the market do not justify *both* a considerable number of new and modernised mines *and* the retention of a large number of older mines on all the coalfields of the country. Hard choices will have to be made.

Such a reappraisal will not mean the total abandonment of the coal industry's modernisation and investment programme. It *is* likely, however, to require some alteration in the scale and the timing, and perhaps even the geography, of the industry's plans. Only with such a change will the coal industry properly serve the national economic interest. A failure to meet the challenge, on the other hand, would leave the British coal industry unnecessarily weakened and the public purse considerably the poorer.

Bibliography

Berkovitch, I. 1977. *Coal on the switchback – the coal industry since nationalisation*. London: George Allen & Unwin.

Brodman, J. R. and R. E. Hamilton 1979. *Comparisons of energy projections to 1985*. Paris: International Energy Agency, OECD.

Bureau of Mines 1976. *Basic estimated capital investment and operating costs for coal strip mines*. Washington, DC: US Government Printing Office.

CEGB (Central Electricity Generating Board) 1978a. *Corporate plan 1978*. London: CEGB.

CEGB 1978b. Evidence of J. A. Jukes to House of Lords. *Report on coal*, Select Committee on the European Communities, Session 1978–9, 1st Report, 86–102. London: HMSO.

CEGB 1979. *Vale of Belvoir inquiry document no. 230*. London: CEGB.

CEGB 1980a. *Vale of Belvoir inquiry document no. 264*. London: CEGB.

CEGB 1980b. *Vale of Belvoir inquiry document no. 336*. London: CEGB.

CEGB 1980c. *Vale of Belvoir inquiry document no. 368*. London: CEGB.

CEGB 1980d. *Vale of Belvoir inquiry document no. 405*. London: CEGB.

Commission of the European Communities 1979a. *In favour of an energy efficient society*. Brussels: EEC.

Commission of the European Communities 1979b. *Energy programme of the European Communities* (COM(79)527). Brussels: EEC.

Commission of the European Communities 1980. *Outlook for the long-term coal supply and demand trend in the Community* (COM(80)117). Brussels: EEC.

Cook, P. L. and A. J. Surrey 1977. *Energy policy – strategies for uncertainty*. London: Martin Robertson.

Davies, D. 1978. Report of a speech, quoted by Stocks (1980), day 56, p. 82.

DEn (Department of Energy) 1976). *Report of the working group on energy elasticities*. Energy paper no. 17. London: HMSO.

DEn 1977. *Energy policy review*. Energy paper no. 22. London: HMSO.

DEn 1978a. *Energy forecasting methodology*. Energy paper no. 29. London: HMSO.

DEn 1978b. *Coal and nuclear power station costs*. Energy Commission paper no. 6. London: Department of Energy.

DEn 1979a. *Energy projections 1979*. London: Department of Energy.

DEn 1979b. *Energy conservation: scope for new measures and long-term strategy*. Energy paper no. 33. London: HMSO.

DEn 1980a. *Energy Trends*, monthly. London: Department of Energy.

DEn 1980b. *Vale of Belvoir inquiry document No. 253*. London: HMSO.

DEn 1980c. Vale of Belvoir inquiry, cross-examination of T. P. Jones, days 29 and 30.

Electricity Council 1979. *Medium term development plan, 1979–86.* London: Electricity Council.

Fernie, J. 1980. *A geography of energy in the United Kingdom.* London: Longman.

IEA (International Energy Agency) 1978. *The electricity supply industry 1974/76 and prospects to 1980/1985/1990.* Paris: OECD.
IEA 1979. *Steam coal prospects to 2000.* Paris: OECD.
Industrial Research Institute 1976. *Technical and economic study on the availability of coal, nuclear and new energy.* Tokyo: Government Publishing Company.

Jevons, S. 1865. *The coal question.* London: Macmillan.

Leach G. *et al.* 1979. *A low energy strategy for the United Kingdom.* London: International Institute for Environment and Development.
London Borough of Newham 1978. *Beckton District Plan; statement of evidence by the British Gas Corporation* (Objection no. 30). Department of Planning and Architecture, LBN.
Lucas, N. J. D. 1977. *Energy and the European Communities.* London: Europa.

Mann, C. E. and J. N. Heller 1978. *Coal and profitability: an investor's guide.* New York: McGraw-Hill.
Manners, G. 1976. The changing energy situation in Britain. *Geography* **61** (4), 221–31.
Manners, G. 1978. Alternative strategies for the British coal industry. *Geog. J.* **144**(2), 224–34.
Manners, G. 1980a. Vale of Belvoir Inquiry, proof of evidence, days 67 and 68.
Manners, G. 1980b. Prospects and problems for an increased resort to coal in Western Europe, 1980–2000. In *Selected studies on energy*, H. H. Landsberg (ed.). Cambridge, Mass: Ballinger.
Manners, G. 1980c. Prospects and problems for an increased resort to nuclear power in Western Europe, 1980–2000. In *Selected studies on energy*, H. H. Landsberg, (ed.). Cambridge, Mass: Ballinger.
Minister of State for Energy 1980. Statement to House of Commons, *Hansard*, 25 February, Col. 1023.
Moses, K. 1979. Vale of Belvoir Inquiry, proof of evidence and cross examination, days 9 and 10.
Muir, W. L. G. 1975. *Review of the world coal industry to 1990.* Wembley: Miller Freeman.

NCB (National Coal Board) 1974. *Plan for coal.* London: NCB.
NCB 1979. Submission to the Commission on Energy and the Environment, unpublished.
NCB 1980. *Vale of Belvoir inquiry document no. 402.* London: NCB.
New Statesman 1978. What future for the miners?, 7 July, pp. 8–10.

North, J. and D. Spooner 1977. The great UK coal rush. *Area* 9(1), 15—27.

North, J. and D. Spooner 1978. On the coal-mining frontier. *Town and Country Planning*, March, 155—63.

North, J. and D. Spooner 1979. The geography of the coal industry in the UK in the 1970s. *Geog. J.* 2(3), 255—72.

Opencast Mining Intelligence Group 1979. *A reassessment of opencast coal-mining*. Leeds: OMIG.

Price Commission 1979. *Area electricity boards — electricity prices and custom allied charges*. Report no. 42. London: HMSO.

Razzell, R. N. 1980. Vale of Belvoir Inquiry, proof of evidence and cross examination, days 32 and 33.

Robens, Lord 1972. *Ten year stint*. London: Cassell.

Robinson, C. 1980. Vale of Belvoir Inquiry, proof of evidence, day 40.

Secretary of State for Energy 1978. *Energy policy: a consultative document, Cmnd 7101*. London: HMSO.

Secretary of State for Energy 1979. Statement to House of Commons, *Hansard*, 18 December, Col. 287.

Secretary of State for Energy 1980a. Speech on second reading of the Coal Industry Bill in the House of Commons, *Hansard*, 17 June, Col. 1377.

Secretary of State for Energy 1980b. Speech on third reading of the Coal Industry Bill in the House of Commons, *Hansard*, 24 July, Col. 813.

Shand, A. T. 1978. In M. Portillo (ed.), *National Coal Conference '78*. London: Conservative Central Office.

Stocks, J. 1980. Vale of Belvoir Inquiry, Proof of evidence, days 56 and 57.

Tregelles, P. G. 1976. A typical colliery in the year 2000. *Colliery Guardian Ann. Rev.*, August, pp. 411—16.

Tripartite Group 1977. *Coal for the future*. London: Department of Energy.

White, N. A. 1978. Minerals for energy; the international availability of energy materials. *CIM Bull.*, September, 57—67.

Williams, L. 1980. Vale of Belvoir Inquiry, proof of evidence and cross examination, day 17.

WOCOL (World Coal Study) 1980. *Coal — bridge to the future*. Cambridge, Mass: Ballinger.

Index

9 780367 231224